Energy Conversion and Green Energy Storage

Energy Conversion and Green Energy Storage presents recent developments in renewable energy conversion and green energy storage. Covering technical expansions in renewable energy and applications, energy storage, and solar photovoltaics, this book features chapters written by global experts in the field.

Providing insights related to various forms of renewable energy, this book discusses developments in solar photovoltaic applications.

Features
- Covers developments in thermal energy storage, photovoltaics, and solar energy.
- Discusses phase change materials and solar cell materials.
- Includes results obtained from various simulation programs, such as Wien2k code, VASP code, and MATLAB®.
- Features green-technology solutions for energy storage devices.
- Presents chapters written by global experts in the field.

This book serves as a useful reference for researchers, graduate students, and engineers in the field of energy.

Energy Conversion and Green Energy Storage

Edited by
Amit Soni
Dharmendra Tripathi
Jagrati Sahariya
Kamal Nayan Sharma

CRC Press
Taylor & Francis Group
Boca Raton London New York

CRC Press is an imprint of the
Taylor & Francis Group, an **informa** business

MATLAB® is a trademark of The MathWorks, Inc. and is used with permission. The MathWorks does not warrant the accuracy of the text or exercises in this book. This book's use or discussion of MATLAB® software or related products does not constitute endorsement or sponsorship by The MathWorks of a particular pedagogical approach or particular use of the MATLAB® software.

First edition published 2023
by CRC Press
6000 Broken Sound Parkway NW, Suite 300, Boca Raton, FL 33487-2742

and by CRC Press
4 Park Square, Milton Park, Abingdon, Oxon, OX14 4RN

CRC Press is an imprint of Taylor & Francis Group, LLC

© 2023 selection and editorial matter, Amit Soni, Dharmendra Tripathi, Jagrati Sahariya, and Kamal Nayan Sharma; individual chapters, the contributors

Reasonable efforts have been made to publish reliable data and information, but the author and publisher cannot assume responsibility for the validity of all materials or the consequences of their use. The authors and publishers have attempted to trace the copyright holders of all material reproduced in this publication and apologize to copyright holders if permission to publish in this form has not been obtained. If any copyright material has not been acknowledged please write and let us know so we may rectify in any future reprint.

Except as permitted under U.S. Copyright Law, no part of this book may be reprinted, reproduced, transmitted, or utilized in any form by any electronic, mechanical, or other means, now known or hereafter invented, including photocopying, microfilming, and recording, or in any information storage or retrieval system, without written permission from the publishers.

For permission to photocopy or use material electronically from this work, access www.copyright.com or contact the Copyright Clearance Center, Inc. (CCC), 222 Rosewood Drive, Danvers, MA 01923, 978-750-8400. For works that are not available on CCC please contact mpkbookspermissions@tandf.co.uk

Trademark notice: Product or corporate names may be trademarks or registered trademarks and are used only for identification and explanation without intent to infringe.

ISBN: 978-1-032-19206-2 (hbk)
ISBN: 978-1-032-19220-8 (pbk)
ISBN: 978-1-003-25820-9 (ebk)

DOI: 10.1201/9781003258209

Typeset in Times
by codeMantra

Contents

Preface vii
Acknowledgments ix
Editors xi
Contributors xv

PART I Solar Photovoltaic Emerging Technologies

Chapter 1 Photocatalyst: Potential Materials for Energy Production and Conversion 3

    B.R. Bhagat and Alpa Dashora

Chapter 2 Design and Development of IoT-Based PV Cleaning System 29

    Vinay Gupta, Amit Soni, Himanshu Priyadarshi, and Ronit Banerjee

Chapter 3 End Life Cycle Cost-Benefit Analysis of 160 kW Grid-Integrated Solar Power Plant: BSDU Jaipur Campus 43

    Pancham Kumar, Manisha Sheoran, Anupam Agrawal, Amit Soni, and Jagrati Sahariya

Chapter 4 A Review of the Theoretical Results Associated with the Intermediate Bandgap Solar Cell Materials: A Density Functional Study 57

    Aditi Gaur, Karina Khan, Amit Soni, Jagrati Sahariya, and Alpa Dashora

Chapter 5 Finite Volume Numerical Analysis of Diamond and Zinc Nanoparticles Performance in a Water-Based Trapezium Direct Absorber Solar Collector with Buoyancy Effects 77

    Sireetorn Kuharat, O. Anwar Bég, Henry J. Leonard, Ali Kadir, Walid S. Jouri, Tasveer A. Bég, B. Vasu, J.C. Umavathi, and R.S.R. Gorla

Chapter 6 Thermal Performance Study of a Copper U-Tube-based Evacuated Tube Solar Water Heater 101

    Arun Uniyal and Yogesh K. Prajapati

PART II Green Energy Storage

Chapter 7 Green Technology Solutions for Energy Storage Devices 117

Himanshu Priyadarshi, Kulwant Singh, and Ashish Shrivastava

Chapter 8 Computational Fluid Dynamic Simulation of Thermal Convection in Green Fuel Cells with Finite Volume and Lattice Boltzmann Methods 133

O. Anwar Bég, Hamza Javaid, Sireetorn Kuharat, Ali Kadir, Henry J. Leonard, Walid S. Jouri, Tasveer A. Bég, V.R. Prasad, Z. Ozturk, and Umar F. Khan

Chapter 9 Graphene-based Composites for High-Speed Energy Storage Battery Application 171

Satya Narayan Agarwal, Ashish Shrivastava, and Kulwant Singh

Chapter 10 Energy Applications of Ionic Liquids 185

Moumita Saha, Manoj K. Banjare, Kamal Nayan Sharma, Gyandshwar K. Rao, Anirban Das, Monika Vats, Gaurav Choudhary, Kamalakanta Behera, and Shruti Trivedi

Index 211

Preface

Global energy demand has exponentially increased due to irrepressible factors such as growing population, industrialization, and fast depleting fossil fuels. Massive utilization of fossil fuels such as coal, oil, and gas for desired fulfillment of energy demand has resulted in an increase in pollution worldwide and in serious human health issues. Renewable sources employed in recent years have exceptional potential and hence resulted in feasible solutions for fulfilling the raised energy demand. Among available renewables, green nanotechnology implemented using nanomaterials has reflected promising results without harming the environment or human health. Green nanotechnology employs non-toxic ingredients that are efficient and environmentally friendly and generates fewer waste products. *Energy Conversion and Green Energy Storage* presents chapters associated with recent developments in renewable energy conversion including green nanotechnology as one of the major initiatives. This will help to throw light on cutting-edge research and will be a ready reference for researchers working in this domain. In this book, the main emphasis is given to technical expansions linked with renewable energy, nanotechnology applications, energy storage, solar photovoltaic, and various other related areas. Readers will find *Energy Conversion and Green Energy Storage* as a useful research contribution and can also refer to the textbook for courses taken by graduate and undergraduate students. In particular, this book will provide insights related to various forms of renewable energy, such as solar photovoltaic applications, energy storage, nanomaterials, nanofluids, and CFD simulations of various aspects of nanofluids/hybrid nanofluids. This book presents a very useful and readable collection of review and research papers in the field of nanotechnologies for energy conversion, storage, and utilization, offering new results which will be of interest to researchers, students, and engineers in the field of nanotechnologies and energy.

The objectives of this book are subsequently presented as follows:

Chapter 1 presents the current scenario of energy and environmental issues, in which solar energy is presented as a viable renewable energy source that can replace fossil fuels on a massive scale.

Chapter 2 elaborates that fossil fuels are predicted to be on the verge of exhaustion, coupled with their environmental hazards; a paradigm shift toward renewable energy utilization is being spearheaded by insolation harvesting. This chapter presents the design and development of the IoT-based cleaning system for the PV module.

Chapter 3 is a case study presented on electricity generation from the 160 kW grid-tied solar rooftop power plant installed in BSDU Jaipur campus situated at 26°9124 latitude N and 75°7873 E longitude geographical location.

Chapter 4 reports that among existing renewable energy sources, solar energy is the most abundant, authentic, ecofriendly, and inexhaustible source. Solar cells, especially with intermediate bandgap, attract the attention of researchers due to their high efficiency. This chapter provides a detailed review of the different types of intermediate bandgap solar cells which includes binary (e.g., InAs, GaAs), ternary (e.g., GaNAs, CuGaS$_2$), quaternary (e.g., CnInGaS), and doped materials (e.g., transition metal-doped CuGaS$_2$).

Chapter 5 deals with nanomaterials that have been deployed in solar energy systems such as carbon-based (e.g., silicates, diamond, carbon nanotubes) and metallic nanoparticles (e.g., gold, silver, copper, tin, zinc). The combination of these nanoparticles with water base fluids will create nanofluids, resulting in improved performance in direct absorber solar collector (DASC) systems. This chapter presents the simulation of carbon-based and metal-based nanoparticles in a trapezium geometry by employing finite volume code (ANSYS FLUENT ver. 19.1).

Chapter 6 presents an experimental work that is carried out to investigate the heat transfer performance and thermal efficiency of a U-tube-based evacuated tube solar water heater.

Chapter 7 presents a critical survey of recent advances in the field of green-nanotechnology approach for achieving improved metrics in energy storage devices.

Chapter 8 presents the modern thrust for green energy technologies based on considerable efforts in developing efficient, environmentally friendly fuel cells. In this chapter, the current fuel revolution which is a reason for the implementation of several hybrid vehicles commercially is presented.

Chapter 9 indicates that, in recent years, carbon and its derivatives have been given much attention by the scientific community for energy storage devices. In this chapter, the recent advances in graphene-based nanocomposites such as SnO$_2$, Mn$_3$O$_4$, Co$_3$O$_4$, and Ti$_5$O$_{12}$, which have shown promising anode materials for Li-ion batteries, are discussed in detail. All the nanocomposites exhibit additional capacity relative to the pure graphene, which is to be expected given the intercalation capability of each of the compounds. This chapter further highlights the working principles and problems hindering the practical applications of graphene-based materials in lithium batteries, supercapacitors, and fuel cells. Future research trends toward new methodologies to the design and the synthesis of graphene-based nanocomposite with unique architectures for energy storage devices and conversion are also presented.

Chapter 10 studies the ionic liquids (ILs) that are currently a topic of immense importance due to their unique properties and widespread application potential. In this chapter, various energy applications of ILs are presented, with a main focus on their utility in fuel cells, supercapacitors, batteries, and dye-sensitized solar cells. Due to their characteristic properties such as low volatility coupled with high electrochemical and thermal stability, as well as high ionic conductivity, ILs appear to meet the rigorous demands/criteria of these various energy applications.

Acknowledgments

Energy Conversion and Green Energy Storage is one among the available unique contributions in relation to contemporary developments occurring in the field of renewable energy conversions and green nanotechnology. This book includes research as well as review articles which are collated here with the support from the research community actively working in the field of solar energy conversion, storage, and green nanotechnology.

The present effort could not be accomplished without the unconditional support provided by many individuals who have contributed their expertise, effort, and encouragement in all possible manners. We are grateful to all our friends, colleagues, researchers, and administration for their continuous encouragement related to collation of contents which has helped in the timely completion of this book.

We would like to express our sincere gratitude to the leadership of Manipal University Jaipur, Jaipur (Rajasthan), National Institute of Technology, Uttarakhand, Srinagar (Garhwal), Uttarakhand, and Amity University Haryana, Manesar, Gurugram, for extending their exhaustive support that has helped in bringing academic and research communities together at both national and international levels.

Our special thanks to all contributors who have actively participated and submitted detailed chapters in the suggested format and have also responded to all suggested corrections in a timely way, which have helped in fruitful completion of this book. We express our heartfelt thanks to all reviewers of submitted book chapters for their patience in reading and providing valuable feedback for the assigned chapters, which has enhanced the quality of the submitted manuscripts.

Dr. Kyra Lindholm and team members of Taylor & Francis Group deserve special thanks for analyzing our proposal and giving us the opportunity to contribute, which is the sole reason for this finalized version available to all.

Lastly, we sincerely acknowledge loving support, patience, and encouragement from our family members who helped us directly and indirectly in timely completion of this work.

Dr. Amit Soni

Dr. Dharmendra Tripathi

Dr. Jagrati Sahariya

Dr. Kamal Nayan Sharma

Editors

Dr. Amit Soni, Professor, Department of Electrical Engineering, Manipal University Jaipur, Jaipur (Rajasthan), India.

Prof (Dr.) Amit Soni graduated in Electrical (Electronics & Power) Engineering from SRTMU, Maharashtra, in 2001 and was awarded 'Gold Medal' for his outstanding academic performance. He completed his MTech (Power System) in 2005 and PhD (Power System) in 2012 – both at Malaviya National Institute of Technology (MNIT), Jaipur.

Dr. Amit Soni joined Manipal University Jaipur on August 20, 2014, and is currently working as Director (International Collaborations) since October 1, 2021. In addition to the current position, he has served as Director (Quality & Compliance) from July 1, 2020, to September 2020 and successfully contributed for NIRF Ranking 2021, AQAR-1, and NBA visits for five Engineering Departments. He has rich working experience of 20 years from various institutions, which includes 3 years of industrial experience in RRVPNL, Jaipur. Before joining Manipal University Jaipur, he has performed various administrative roles such as Director, Asians Institute of Technology, Rajasthan; Head, Electrical Department; and Coordinator and Chairman for various academic committees at both undergraduate and postgraduate levels.

He has served as Professor & Head, Department of Electrical Engineering for 4.5 years from December 18, 2015, and during this tenure, he has successfully organized several national and international workshops and Scopus-indexed international conferences, initiated MoUs and also developed research laboratories.

Prof. Amit Soni is currently working in interdisciplinary research areas which include solar photovoltaic materials, optoelectronics, thin-film technology, renewable energy systems, and power system. He has published 10 Scopus-indexed book chapters and 70 research papers in reputed SCI-indexed international journals and Scopus-indexed conferences which include high-impact Q1 & Q2 journals of repute. He has delivered several invited talks, keynote addresses, and chaired/co-chaired sessions in various reputed indexed conferences. He is a regular reviewer of research articles for many high-impact journals. He has successfully supervised three PhD students, and six research scholars are working under his supervision. He is currently working as PI for DST SERB-funded research project in collaboration with MLSU, Udaipur and NIT, Uttarakhand. He is a life member of the Solar Energy Society of India, and a member of ISTE, India and IEEE, USA.

Dr. Dharmendra Tripathi, Associate Professor, Department of Mathematics, National Institute of Technology Uttarakhand, Srinagar (Garhwal), India.

Dr. Dharmendra Tripathi is an Associate Professor in the Department of Mathematics, National Institute of Technology, Uttarakhand. Prior to joining NIT Uttarakhand, he has worked for more than ten years as a faculty member (Associate Professor, Assistant Professor) in various reputed institutions such as

Manipal University Jaipur, NIT Delhi, IIT Ropar, and BITS Pilani Hyderabad. He has completed his PhD in Applied Mathematics (Mathematical Modeling of Physiological Flows) in 2009 at the Indian Institute of Technology BHU and MSc in Mathematics at Banaras Hindu University.

Dr. Tripathi has supervised six PhD students and three are working under his supervision. He has also guided 20 B.Tech. projects. He has published more than 150 papers in reputed international journals, 1 book with Springer, and 5 book chapters, and presented more than 40 papers in international and national conferences. He has delivered more than 50 lectures as invited speaker, keynote speaker, and resource person in various conferences, workshops, FDP, STTP, STC, refresher courses, etc. His research h-index is 40 and i-10 index is 121, and his papers have more than 5,000 citations.

He has been listed among top researchers/scientists across the world as per updated science-wide author databases of standardized citation indicators in 2020 and 2021. Dr. Tripathi has received President Award in 2017 from the Manipal University Jaipur for outstanding contribution. He has been recognized by the Head of Institution for excellent work and contribution for the NIT Uttarakhand and also recognized by various reputed journals for reviewing the articles and editing the special issues for the journals. He was awarded some prestigious fellowships such as the INAE fellowship in 2015, 2016, 2017 and 2018 and postdoctoral fellowships (NBHM, Dr. D.S. Kothari and Indo-EU) in 2010. He has also organized various events such as national and international conferences/STC/STTP/FDP/workshops/winter schools on various emerging topics. He is a lifetime member of various professional bodies, member of the editorial board of two journals, and reviewer of more than 50 international journals.

Dr. Tripathi has been discharging additional administrative responsibilities as Dean (R&C) at NIT Uttarakhand since June 2019, and he has discharged many administrative responsibilities of the Institutes including I/c Registrar, CVO, Dean (SW), Chief Warden, and Chairman of various institutes' committees.

His research work is focused on the mathematical modeling and simulation of biological flows in deformable domains, peristaltic flow of Newtonian and non-Newtonian fluids, dynamics of various infectious diseases, microfluidics, CFD, biomechanics, heat transfer, nanofluids, energy systems, numerical methods, etc.

Dr. Jagrati Sahariya, Assistant Professor, Department of Physics, National Institute of Technology Uttarakhand, Srinagar (Garhwal), India.

Dr. Jagrati Sahariya has been working in the field of γ-ray scattering and band structure calculations for the last 10 years. She earned her PhD in 2012 in the field of electronic structure calculation and Compton scattering. As part of her PhD thesis, she studied electronic and magnetic properties of a variety of technologically important materials. Dr. Sahariya has sufficient expertise in using different band structure methods such as full-potential linearized augmented plane wave, linear combinations of atomic orbitals and spin-polarized relativistic Korringa–Kohn–Rostoker to compute electronic structure, optical and magnetic properties, and Compton profiles of a variety of materials. She has published more than 70 papers in peer-reviewed international journals of high-impact factor and reputed conferences both at national and international levels. She has also executed a research project funded by SERB under the Fast Track Scheme.

Dr. Kamal Nayan Sharma, Assistant Professor, Department of Chemistry, Biochemistry & Forensic Science, Amity School of Applied Sciences, Amity University Haryana, India.

   Dr. Kamal Nayan Sharma is an assistant professor in the Department of Chemistry, Amity University Haryana, Manesar, Gurugram. Prior to joining Amity University, he has worked as an assistant professor at Vivekanand Global University Jaipur, Rajasthan, India. After earning his PhD from IIT, Delhi, he executed a young scientist project funded by SERB, New Delhi, India, at MNIT Jaipur as a Principal Investigator. He has published 26 papers in international journals of high-impact factors, such as *Nanoscale, Organometallics, Chemical Communications, Dalton Transaction, Organic & Biomolecular Chemistry*, and *Tetrahedron Letters*. His research h-index is 12 and i-10 index is 14, and his papers have more than 540 citations. Dr. Sharma has visited the University of Cape Town, South Africa, for research work under the Department of Science and Technology (Govt. of India)–sponsored Indo-South African Research project. He has also served as a reviewer in various international journals. He is guiding one student for her PhD and guided several Master's students for their project work.

Contributors

Satya Narayan Agarwal
Department of Electrical Engineering
Manipal University Jaipur
Jaipur, India

Anupam Agrawal
Department of Electrical Engineering
Rungta College of Engineering and Technology
Bhilai, India

Ronit Banerjee
Department of Electrical Engineering
Manipal University Jaipur
Jaipur, India

Manoj K. Banjare
MATS School of Sciences
MATS University
Raipur, India

O. Anwar Bég
Aeronautical & Mechanical Engineering Department
School of Science, Engineering and Environment (SEE)
University of Salford
Manchester, United Kingdom

Tasveer A. Bég
Engineering Mechanics Research
Israfil House
Manchester, United Kingdom

Kamalakanta Behera
Department of Applied Chemistry (CBFS - ASAS)
Amity University
Gurugram, India

B.R. Bhagat
Computational Material Science Laboratory, Department of Physics,
Faculty of Science,
The M S University of Baroda
Vadodara, Gujarat, India

Gourav Chaudhary
Department of Applied Chemistry (CBFS - ASAS)
Amity University
Gurugram, India

Anirban Das
Department of Applied Chemistry (CBFS - ASAS)
Amity University
Gurugram, India

Alpa Dashora
Computational Material Science Laboratory
Department of Physics
Faculty of Science
The M S University of Baroda
Vadodara, Gujarat, India

Aditi Gaur
Department of Electrical Engineering
Manipal University Jaipur
Jaipur, India

R.S.R. Gorla
Department of Aeronautics and Astronautics
US Air Force Institute of Technology
Wright-Patterson Air Force Base
Ohio, USA

Vinay Gupta
Department of Electrical Engineering
Manipal University Jaipur
Jaipur, India

Hamza Javaid
Department of Mechanical and
  Aeronautical Engineering
School of Science, Engineering and
  Environment (SEE)
University of Salford
Manchester, United Kingdom

Walid S. Jouri
Department of Mechanical and
  Aeronautical Engineering
School of Science, Engineering and
  Environment (SEE)
University of Salford
Manchester, United Kingdom

Ali Kadir
Department of Mechanical and
  Aeronautical Engineering
School of Science, Engineering and
  Environment (SEE)
University of Salford
Manchester, United Kingdom

Karina Khan
Department of Physics
Manipal University Jaipur
Jaipur, India

Umar F. Khan
Electromagnetics and Fuel Cell
  Research
School of Engineering
Robert Gordon University
Aberdeen, Scotland

Sireetorn Kuharat
Department of Mechanical and
  Aeronautical Engineering
School of Science, Engineering and
  Environment (SEE)
University of Salford
Manchester, United Kingdom

Pancham Kumar
School of Electrical Skills
Bhartiya Skill Development University
Jaipur, India

Henry J. Leonard
Department of Mechanical and
  Aeronautical Engineering
School of Science, Engineering and
  Environment (SEE)
University of Salford
Manchester, United Kingdom

Z. Ozturk
Energy Modelling
Think Lab
University of Salford
Manchester, United Kingdom

Yogesh K. Prajapati
Department of Mechanical Engineering
National Institute of Technology
Uttarakhand, Srinagar, India

V.R. Prasad
Department of Mathematics
School of Advanced Sciences
Vellore Institute of Technology
Vellore, India

Himanshu Priyadarshi
School of Electrical, Electronics, and
  Communication Engineering
Manipal University Jaipur
Jaipur, India

Contributors

Gyandshwar K. Rao
Department of Applied Chemistry
  (CBFS – ASAS)
Amity University
Gurugram, India

Moumita Saha
Department of Chemistry
Institute of Science
Banaras Hindu University
Varanasi, India

Jagrati Sahariya
Department of Physics
National Institute of Technology,
  Uttarakhand
Srinagar, India

Kamal Nayan Sharma
Department of Applied Chemistry
  (CBFS – ASAS)
Amity University
Gurugram, India

Manisha Sheoran
Bhartiya Skill Development University
Jaipur, India

Ashish Shrivastava
Department of Electrical Engineering
Manipal University Jaipur
Jaipur, India

Kulwant Singh
Department of Electronics and
  Communication Engineering
Manipal University Jaipur
Jaipur, India

Amit Soni
Department of Electrical Engineering
Manipal University Jaipur
Jaipur, India

Shruti Trivedi
Department of Chemistry
Institute of Science
Banaras Hindu University
Varanasi, India

J.C. Umavathi
Department of Mathematics
Gulbarga University
Gulbarga, Karnataka, India

Arun Uniyal
Department of Mechanical Engineering
National Institute of Technology
Uttarakhand, Srinagar, India

B. Vasu
Department of Mathematics
MNNIT
Prayagraj, Uttar Pradesh, India

Monika Vats
Department of Applied Chemistry
  (CBFS – ASAS)
Amity University
Gurugram, India

Part I

Solar Photovoltaic Emerging Technologies

1 Photocatalyst
Potential Materials for Energy Production and Conversion

B.R. Bhagat and Alpa Dashora

CONTENTS

1.1 Introduction 3
1.2 Photocatalytic Mechanism 5
    1.2.1 Hydrogen Evolution Reaction 7
    1.2.2 Oxygen Evolution Reaction 8
    1.2.3 Carbon Dioxide Reduction Reaction 9
1.3 Functionalization Methods for Photocatalytic Activity Enhancement 11
    1.3.1 Anionic and Cationic Doping 11
    1.3.2 Co-doping 14
    1.3.3 Semiconductor Heterostructure and Metallic Co-catalyst 15
1.4 Role of Charge Transfer in Enhancing Photocatalytic Activity 18
1.5 Conclusion 19
References 19

1.1 INTRODUCTION

Nature has been a source of inspiration for human beings from the beginning of time for mimicking and utilizing various advanced techniques of energy production and conversion at a large scale [1–3]. In the current scenario with energy and environmental issues, several techniques for waste management and energy generation at different levels have been developed, but the problem remains intact and is expected to increase our concern in the near future [4]. In order to overcome this issue, a green, sustainable and low-cost methodology is necessary. In this direction, solar energy seems to be the only viable renewable energy source that is consumed and converted by living species in various forms, where photosynthesis is being widely replicated worldwide in the form of artificial photosynthesis for photocatalytic carbon dioxide reduction reaction (CO_2RR) for transformation of high environment pollutant gas (CO_2, NH_3) into fuels such as methanol or methane and for air purification globally by photocatalytic paints and building materials in high-rising architectures [5–13].

DOI: 10.1201/9781003258209-2

This process is also used for hydrogen production with simultaneous oxygen evolution reactions. Photocatalyst has emerged as a cure to paralyzed conventional water treatment methods that had high operating cost along with its ability to cause secondary pollution [14,15]. It has also helped in degradation of industrial dyes and paints released into the water bodies, water purification and water splitting into fuel production by utilizing metal oxides that prevent photo-corrosion. This is now a multifunctional research and development field for the scientific community across the world [16–23].

Various metal oxide semiconductor thin films and nanoparticles have shown a possible solution as an efficient photocatalyst for remediation and as an alternate for the treatment of contaminated water. On the basis of their stability, high quantum yield, suitable band alignment and large reactive surface, TiO_2 [24–26], WO_3 [27–29], ZnO [30–32], CdS [33,34] and ZnS [35–37] have proved and are being used as potential candidates for photocatalytic applications. Having multiple phases and facets with band gap value in ultraviolet range, TiO_2 has received attention as a material with a wide band gap for cationic and anionic impurity accommodation, which creates metastable states between valence band and conduction band for charge separation [38–40]. On the other hand, polymorphic ZnO has performed well because of its high thermal stability and possibility of functionalization in the degradation of organic pollutants. However, the activity in ultraviolet region and high heavy-metal content limit their usage in visible spectrum of light and for treatment of water to be consumed later by living species. To utilize maximum region of visible spectrum, various approaches have been tried, but the heavy-metal ion issue remains the same [41,42].

As a photocatalyst, alternative to metal oxides, the journey of carbon nitride started from its first synthesis as polymeric melon, reported in 1834 by Berzelius and Liebig [43] as one of the oldest synthetic polymers. Later, Liu and Cohen [44] constructed the beta phase by replacing Si from β-Si_3N_4. From 1989 to 1996, carbon nitride was evolved, synthesized and reported as a multifunctional compound with a total of eight polymorphs each having a separate role. α-C_3N_4, β-C_3N_4, cubic-C_3N_4, pseudocubic-C_3N_4 and graphitic carbon nitride, which have an s-triazine-based hexagonal structure, s-triazine-based orthorhombic structure and tri-s-triazine-based structure also known as heptazine [45–50], are different allotropes, while tri-s-triazine-based g-C_3N_4 is found to be the most stable structure. g-C_3N_4 structure is a nitrogen heteroatom-substituted graphite framework which includes π-conjugated graphitic planes and sp^2 hybridization of carbon and nitrogen atoms along with numerous photo-reactive and photo-responsive sites. It is a widely known photocatalyst with good visible light absorption, a multi-layered porous structure, great stability and non-toxicity [51–54], but due to high charge carrier (e^-–h^+ pair) recombination tendency, overall photocatalytic efficiency reduces significantly. Many modifications have been included to increase its activity in the visible region of spectrum along with decreasing the rate of recombination of photogenerated charge carriers for enhanced photoactivity [55–58].

This chapter voyages through the functionalization of g-C_3N_4 undertaken to increase its photo-responsive properties in terms of catalysis and presents a comparative study based on various modification strategies. Structural modification of bulk form into nanosheets, nanoflakes, nanotubes and nanoparticles along with modification from planar to corrugated geometry and formation of bi-, tri- and four-layered stacking has been seen for reducing the recombination rate of photogenerated charge carriers and increasing the reactive surface [59–70]. Cationic and anionic doping is a frequently observed technique to reduce the band gap for higher absorption. Doping of several electron-rich elements has caused the formation of deep trap sites, also called intermediate band for reducing recombination rate of photogenerated e−−h+ pair. Nobel metal inclusion shows local surface plasmon resonance enhancing the charge separation leading to higher reactivity.

However, simultaneous reduction–oxidation reaction is difficult to achieve from a single material. Thus, synergistic effects with band mismatch method have been utilized to create a multifunctional catalyst with high absorption by formation of heterostructures via various schemes for efficient charge transfer. At last, this chapter provides deep insight into the development of environmentally friendly carbon nitride photocatalyst as a potential candidate by systematically comparing reported efficiencies based on modification methods and their future scope for charge facilitation over and within layers [71–81].

1.2 PHOTOCATALYTIC MECHANISM

Synergistic effect where photons interact with the photo-responsive material and accelerate the chemical reaction is the prima step of photocatalysis. On absorbing the photon of energy equivalent or greater than band gap of a semiconductor catalyst, electron excites from the valence band (VB) states to the conduction band (CB) states resulting in the formation of a photogenerated electron–hole pair in CB–VB. These charge carriers migrate to the reaction site to participate in the corresponding reaction where the electrons take part in the reduction reaction forming hydrogen molecule, superoxide anions by reducing the ambient oxygen, $HCOO^-$ via carbon dioxide reduction and ammonia production through nitrogen reduction, while holes oxidize adsorbed water molecule in the form of moisture from the atmosphere to generate hydroxyl radicals which are useful for oxygen evolution reaction and dye degradation [82–86]. However, low band gap value, high absorption and large photogenerated electron–hole pair satisfy the essential step for photocatalysis, but reduction–oxidation process depends on band-edge positions of the material. Straddling redox potential is required to provide reduction–oxidation environment to the photocatalyst which includes conduction band minima (CBM) to be higher than the H_2O/H_2 level known as reduction potential (−4.44 eV) and valence band maxima (VBM) to be lower than O_2/H_2O level called oxidation potential (−5.67 eV) against the vacuum potential. Thus, the minimum band gap requirement has to be greater than 1.23 eV for overall water splitting [63,87]. The above process can also be explained with Figure 1.1 in terms of redox potential: band-edge positions like VBM and CBM are determined with respect to normal hydrogen electrode potentials using work function (ϕ) and band gap (E_g) as follows:

$$\text{Conduction band edge } (E_{CB}) = \phi - 4.44\,\text{eV} - E_g \tag{1.1}$$

$$\text{Valence band edge}(E_{VB}) = \phi - 4.44\,\text{eV} + E_g \tag{1.2}$$

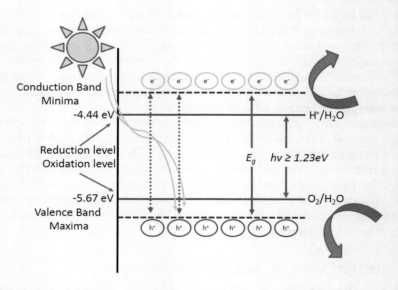

FIGURE 1.1 Photocatalysis mechanism with reduction–oxidation potential and valence-conduction levels. (Energy scale is with respect to vacuum energy.)

For bulk materials, absolute electronegativity (χ) is considered in place of ϕ in Eqs. (1.1) and (1.2). Figure 1.1 shows the schematics of the redox reaction as a part of photocatalytic mechanism. The possibility of a redox reaction is based on the band-edge position, but these band edges are determining the potential for further reaction. Therefore, the potential barrier required for all steps to proceed swiftly should be smaller than the band edges values.

Thus, for the overall water-splitting reaction, four parameters need to be satisfied: (i) high absorption in the visible region, (ii) low charge carrier recombination for high reactivity, (iii) suitable band edges for reduction and oxidation reactions and (iv) reaction energy barrier within the band-edge potential for respective reactions to take place smoothly. The material fulfilling all these four criteria is considered an efficient photocatalyst [88,89]. For the successful screening and validation of above-mentioned criterion for the efficient photocatalyst using first-principles method, a detailed analysis of (i) electronic, (ii) optical and (iii) photocatalytic/thermodynamic properties should be considered as shown in Figure 1.2.

Nowadays, design of an effective photocatalyst using density functional theory (DFT) is one of the most popular, promising, and cost-effective techniques. The required four parameters for efficient photocatalyst can be studied by means of density of states, band structure, ϕ, charge distribution using Löwdin charge transfer and optical property analysis. The number of states present in the system described by the density of states gives the information about the carrier concentration and their energy distribution. Direct and indirect band gap values along with the curvature of the bands are provided by band structure that helps to calculate effective mass (m^*) of the carriers using parabolic fitting method to understand the recombination rate

Steps to Design Novel Photocatalyst

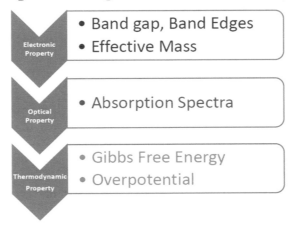

FIGURE 1.2 Stepwise screening methods for efficient photocatalyst.

of electron–hole pair in CBM and VBM. The effective mass of the carrier is defined as follows:

$$m^* = h^2[(\partial^2 E(k))/(\partial k^2)]^{-1} \tag{1.3}$$

The mobility of charge carriers and their effective mass exhibit inversely proportional relation as follows:

$$v = q\tau/m^* \tag{1.4}$$

Here, v: mobility of carriers, q: charge carrier and τ: scattering time [90].

The reaction mechanism can also be understood with the help of DFT. The steps of photocatalytic reaction for reduction and oxidation of various molecules are as follows.

1.2.1 Hydrogen Evolution Reaction

The hydrogen evolution reaction (HER), a part of water-splitting mechanisms, is described by the Volmer–Heyrovsky and Volmer–Tafel reactions shown in Figure 1.3. Volmer–Heyrovsky is energetically more stable than the two-step Volmer–Tafel reaction, where in the former reaction, the electron–hole pair over the substrate (*) forms a hydrogen intermediate (H*) which converts into a hydrogen molecule on interacting with another pair of charge carrier. The role of the water molecule along with the isolated H⁺ is carried out with structural relaxation that describes its interaction and effect on the free energy. Computationally, it includes the formation of hydronium molecule (H_3O) as an intermediate adsorbed over the surface elaborated in the following equations [91,92]:

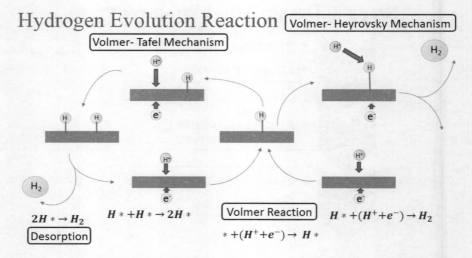

FIGURE 1.3 Hydrogen evolution reaction following the Volmer–Tafel–Heyrovsky mechanism accompanied with desorption.

$$* + (H^+ + e^-) \to H* \tag{1.5}$$

$$H* + (H^+ + e^-) \to * + H_2(g) \tag{1.6}$$

And with inclusion of extra water molecule:

$$H_2O* + (H^+ + e^-) \to H_3O* \tag{1.7}$$

$$H_3O* + (H^+ + e^-) \to H_2O* + H_2(g) \tag{1.8}$$

1.2.2 Oxygen Evolution Reaction

Oxygen evolution reaction (OER) is divided into four one-electron transfer steps as proposed by Rossmeisl et al. [93] and mentioned in Eqs. (1.9–1.12). Here, each step forms an intermediate species (with an asterisk) along with electron–hole pairs, and as the reaction proceeds, there is requirement/involvement of two water molecules for the generation of oxygen molecule releasing a sum of four electrons. OER and HER could simultaneously take place as the released electron–hole pair from the OER can be utilized for generation of two hydrogen molecules. The schematic diagram for OER is also presented in Figure 1.4 for Co–B–C_3N_4 [94].

$$* + 2H_2O(l) \to OH* + H_2O(l) + (H^+ + e^-) \tag{1.9}$$

Photocatalyst

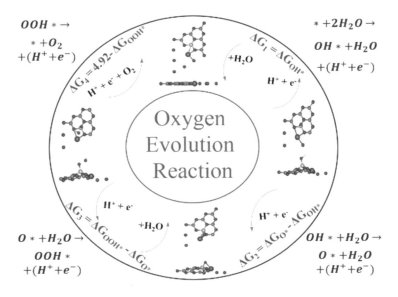

FIGURE 1.4 Four electron step mechanism of oxygen evolution reaction over B-doped-Co-loaded g-C_3N_4 [94].

$$OH * + H_2O(l) \rightarrow O * + H_2O(l) + (H^+ + e^-) \quad (1.10)$$

$$O * + H_2O(l) \rightarrow OOH * + (H^+ + e^-) \quad (1.11)$$

$$OOH * \rightarrow * + O_2(g) + (H^+ + e^-) \quad (1.12)$$

1.2.3 Carbon Dioxide Reduction Reaction

Reactions for carbon dioxide reduction involve eight electron steps. There is a possibility of formation of several unstable intermediates apart from that shown in the reactions based on the substrate used. The simultaneous OER and CO_2RR are suitable, leading to the generation of oxygen to carbon dioxide molecule in 2:1 proportion. OER along with HER/CO_2RR is preferred over one-sided reaction since it prevents accumulation of charge carriers over the surface and decreases the possibility of photo-corrosion [6,95]. The following reactions describe the stepwise conversion of carbon dioxide into various forms of fuel:

$$2CO_2 + e^- \rightarrow CO_2^- \quad (1.13)$$

$$CO_2(g) + H_2O(l) + 2e^- \rightarrow HCOO^-(aq) + 2OH^-(aq) \quad (1.14)$$

$$CO_2(g) + H_2O(l) + 2e^- \rightarrow CO(g) + 2OH^-(aq) \tag{1.15}$$

$$CO_2(g) + 3H_2O(l) + 4e^- \rightarrow HCHO^-(l) + 4OH^-(aq) \tag{1.16}$$

$$CO_2(g) + 5H_2O(l) + 6e^- \rightarrow CH_3OH(g) + 6OH^-(aq) \tag{1.17}$$

$$CO_2(g) + 6H_2O(l) + 8e^- \rightarrow CH_4(g) + 8OH^-(aq) \tag{1.18}$$

Change in Gibbs free energy (ΔG) form calculation of energy barrier for each adsorption reaction is evaluated as follows:

$$\Delta G = \Delta E + \Delta ZPE - T\Delta S + \Delta G_{pH} + \Delta G_U \tag{1.19}$$

Here, ΔE is the change in the total energy evaluated from the DFT study, ΔZPE is the zero-point energy change, T is the temperature, ΔS is the entropy change, $\Delta G_{pH} = -k_B T \ln_{10} \times pH$ and $\Delta G_U = -eU$ (U is the potential measured against normalized hydrogen electrode). The calculation of free energy change in the oxidation/reduction reaction is done using the method developed by Nørskov et al. [96]. The steps that have a maximum change in energy are rate-determining ones. The theoretical value of barrier height for OER is calculated as follows:

$$\eta^{OER} = \text{Max}_i[\Delta G_i] / ne \tag{1.20}$$

where i denotes intermediate.

The entropy of adsorbed hydrogen atoms is far lower than the entropy in the gas phase (H_2). Since the ZPE and entropy are not sensitive to the coverage, entropy of H_2 gas at standard conditions [92] is used to calculate entropy capacity. Nørskov et al. [96] proposed overall correction to the equation

$$\Delta G_{H*} = \Delta E_{H*} + 0.24 \tag{1.21}$$

$$\eta^{HER} = [\Delta G_{H*}] / ne \tag{1.22}$$

Photocatalytic energy conversion and dye degradation are measured in terms of quantum yield which is the ratio of reaction rate to the photon absorption rate, although the calculation of absorbed photon energy is difficult due to its scattering. Thus, the efficiency of fuel production is based on its pristine sample or with respect to certain standard material.

The theoretical study of reaction mechanism, adsorption of intermediates, their surface interaction, intermediate interaction among each other and overall effect over the redox capability with the lifetime study of excitons and their role in the inclusion

Photocatalyst

of van der Waals interaction is still lacking, which plays an important role in the photocatalytic activity as well as enlightens the path for experimentalists to understand the role of catalyst [97,98].

Therefore, the complete theoretical study in terms of photocatalyst performance of any materials should include electronic, optical and thermodynamic (reaction mechanisms). In the next section, the attempts to enhance the photocatalytic activity of g-C_3N_4 using theoretical and experimental techniques are reviewed.

1.3 FUNCTIONALIZATION METHODS FOR PHOTOCATALYTIC ACTIVITY ENHANCEMENT

In continuation to the current status in the field of photocatalyst for pollutant degradation, CO_2 reduction and H_2 production, it is seen that several modifications such as surface alteration, non-metal and/or metal doping and heterostructures formation have been performed to achieve high photocatalytic activity of different 2D semiconductors. Still there is a bottleneck to accept this technology at a large scale due to low quantum yield and efficiency which is mainly caused due to low charge separation and transportation.

Very few complete theoretical studies are available in this field. Therefore, band gap engineering of functionalized semiconductors for photocatalytic application is the primary technique to enhance the performance.

1.3.1 ANIONIC AND CATIONIC DOPING

Doping in the 2D semiconductor photocatalyst is done in order to enhance the optical performance by narrowing the band gap for higher absorbance in the visible region. Figure 1.5 shows change in band edges for g-C_3N_4 after different doping. Doping of various non-metal elements has been undertaken on the monolayer of g-C_3N_4 where band gap alteration for H, B, C, O, F, Si, S, P, Cl, As, Se, Br, Te and I with pristine

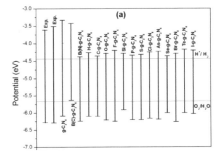
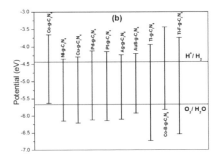

FIGURE 1.5 (a) Band edges for experimental and theoretical pristine and non-metal-doped g-C_3N_4 [94,99]. (b)s Band edges for metal and anion–cation-doped g-C_3N_4 (potential vs. vacuum energy) [94,125].

compounds has been studied [99–103]. Among all the anions, Br and I were found unstable thermally due to large positive adsorption energies, while the Fermi energy of all the samples was reduced except that of I-doped monolayer. All reported samples showed higher CBM than reduction potential and lower value than oxidation potential making them suitable for redox ability. HOMO–LUMO position determines the ability to separate the photogenerated charge carriers which directs towards higher photoactivity [99]. S-doped C_3N_4 reduces the Gibbs free energy by 0.28 eV for CO_2 reduction due to promotion of charge separation studied using HOMO–LUMO positioned at N-S atoms and prolongs charge carrier lifetimes by inhibiting the electron–hole recombination [100,101]. A study on the P-doped system emphasized the activation of π-conjugation after the lamellar exfoliation to readily enhance the photo-response [102]. The multiple charge carrier transfers by proton-coupled transfer mechanism and their effective separation show 1.9, 1.4, 1.7 and 2.4-folds high CH_4 production for O-, P-, B-, and S-doped g-C_3N_4, respectively [103]. Despite limited surface area and photon absorbance of S-doped material, the improvement is subjected to migration and separation of charges. Impurity states formed in the forbidden region have changed the optical absorption along with the obvious redshift after all anionic doping by 10–75 nm [104]. For H-, B-, C-, Si- and Se-integrated ultraviolet region and for all dopants, integrated visible region reduced by 7% and increased by about 14%–71%, respectively. A similar study of S atom or C (N)-vacancy in g-C_3N_4 has expanded the light response range by decreasing the band gap or by producing impurity states [101,105]. Cui et al. [106] and Zhu et al. [107] have reasoned the charge facilitation to delocalized HOMO–LUMO for improved visible absorption for O- and F-doped tri-s-triazine which also modified the reduction–oxidation potential. So far, the work done on non-metal doping has emphasized on the charge separation and π-localization over the surface for higher reactive sites [108].

Photocatalytic properties are enhanced for Co-doped g-C_3N_4 due to formation of Co–N bond [109], whereas Fe and Ni penetration into g-C_3N_4 extends the charge carrier migration distance which suppresses the recombination of electron–hole pairs. On the other hand, Cu and Zn enhance interlayer charge carrier migration [110]. Hussain et al. [111] have designed novel functional nanostructures that are capable of trapping a large number of CO_2 molecules by means of first-principles study. They have functionalized g-C_6N_8 nanosheets with transition metal (TM) dopant from Sc to Zn, and each TM dopant with a doping concentration of 1.79% can anchor maximum of four CO_2 molecules with suitable adsorption energies (−0.15 to −1.0 eV) for ambient condition applications [111].

Bai et al. [112] have adsorbed alkali metal Li, Na, K, Rb and Cs over the heptazine type g-C_3N_4 and have seen metallic nature with absorption peak covering visible light area which shows potential application in optoelectronic devices and for visible light catalysis along with sensor detection application due to the charge transfer from alkali metal to substrate leads in decrement of work function as we go down the alkali metal group [112]. Na, K, Rb and Cs have been studied with intercalation between bilayer form of g-C_3N_4 theoretically and experimentally creating an interlayer bridge for efficient charge transfer and formation of interfacial electric field. K, Rb and Cs decoration increases the oxidation capability benefiting from a decrease in electronic localization and positive shifted VB position, although Na-doping shows reduced photoactivity due to high rate of recombination [113,114]. A study by Jiang et al. [115] provided 3.7 times increased hydrogen production by Na-doped g-C_3N_4 photocatalyst (18.7 μmol/h) to pristine g-C_3N_4 (5.0 μmol/h) [115]. Electro-positivity and ionic radius direct cations not as a substitutional dopant but as a decoration over the surface on the triazine moieties or in the void. Energetically, high Z cations and noble metal prefer void site over the surface of the 2D structure. A study by Nguyen et al. [116] using semi-empirical tight-binding-based method on the influence of Al, Fe, Ag, Mg and Li doping on corrugated g-C_3N_4 showed lower band gap and reduction in vertical ionization potential that raised electron affinity along with Lewis acidity of pristine specimen attributing to charge transfer from metal to g-C_3N_4. Among single metal-doped g-C_3N_4 system, Fe and Ag atoms' wavefunction participates in the HOMO and C/N atom in LUMO directing towards charge transfer for enhancement of catalytic activity [116]. Rhodamine B degradation rate for Mn adsorbed g-C_3N_4 over different concentrations showed threefold increase in photocatalytic efficiency resulting from the electron–hole recombination, formation of impurity states which leads to increase in optical absorption. For Mn doping of 8.06 at.%, band-edge shifting towards reduction potential takes place which is most suitable for enhanced photoactivity. Density functional theory computation confirms the suitability of Mn among all other 3d TM due to its most suitable band straddling and highest binding energy, while the small potential distance between VB edge and oxidation potential makes hole–H_2O interaction easier [117,118]. Fe doping has been known to activate the low coordinated nitrogen for higher adsorption [119,120], and TM embedding results in the facilitation of light absorption by impurity state formation. A similar experimental study by Tonda et al. [121] on Fe-doped g-C_3N_4 emphasized high charge separation, charge transfer and visible light response providing 7 and 4.5 times increased photocatalytic activity by bulk and nanosheets of pristine specimen. A two-step photo-thermal-catalytic process of H_2 and O_2 generation via $2H_2O \rightarrow H_2 + H_2O_2$, $2H_2O_2 \rightarrow 2H_2O + O_2$ performed over Fe/C_3N_4 shows 1.4 μmol for H_2 and 0.5 μmol of O_2 evolution in 12 hours [121]. The apparent quantum efficiency of oxygen generation is found approximately five times the pristine g-C_3N_4 on studying H_2O_2 disproportionation reaction mechanism [122].

Machine learning study utilizing electronegativity and d-band centre of TM atom helped Zhao et al. [123] to screen single-atom catalysts based on adsorption strength of formic acid. Then, energy barrier of reaction was computed for various transition metals such as Sc, Ti, V, Cr, Mn, Fe, Co, Ni, Zr, Mo, Ru, Rh, Pd, Ag, W, Os, Ir, Pt; among these Au@ g-C_3N_4, Rh-, Pd- and Pt@ g-C_3N_4 exhibit thermodynamical stability and kinetic feasibility and also reported for effective formic acid dehydrogenation catalyst which is beneficial for hydrogen transport and storage. Investigation on single Pt metal atom dispersed over g-C_3N_4 monolayer exhibits anisotropic nature. Meanwhile, its experimental synthesis is possible on considering the density of Pt atom less than that of sixfold cavity, thus avoiding atomic clustering [124]. A complete overall water-splitting study was performed by Li et al. [125] utilizing a single metal atom decorated in planar void site for Pt, Cu, Ni, Pd, Ag and Au based on charge redistribution and ionic/covalent bond relaxations. Overpotential values showed relationship with conductive bonding nature which is to improve the HER and OER efficiency. Ni, Cu, Pd and Pt show suppression in recombination rate due to separation of VBM and CBM, while overpotential for Pt and Pd with adsorption over the void and TM decreased approximately by four- and threefold for OER and four- and threefold for HER [125]. In spite of band-edge straddling mentioned by single metal catalyst, effective redistribution of HUMO–LUMO altered the barrier height which is an essential step for determination of complete water-splitting ability of photocatalyst.

1.3.2 Co-doping

Anionic doping enhances the charge migration and facilitates separation, while cationic doping/decoration increases optical absorption by formation of impurity states and plays a vital role in effective mass tuning which leads to suppression of recombination rate of photogenerated charge carriers along with effective charge compensation. The involvement of CBM and VBM together, which is caused by simultaneous metal/non-metal doping, is predicted highly useful for simultaneous oxidation and reduction reaction. A study performed over co-doped g-C_3N_4 by Ma et al. [126] on doping of K and S showed positive VB shifting, the opening of charge transfer channel between two triazine moieties, narrowing of band gap, noncoplanar distribution of band edges and increased overpotential value resulting in enhanced OER attributing to separation of electron–hole pair [126]. Efficient approach towards metal-free co-doping in g-C_3N_4 system leads to the study of S, P co-doped at interstitial and carbon sites by Hu et al. [127] and S, P, O co-doped ultrathin nanosheets by Chu et al. [128] along with Liu et al. [129] have reported increased photocatalytic degradation in the former case and enhanced hydrogen evolution rate reaching 2,480 µmol/h g in the latter case due to suppressed recombination, narrowed forbidden energy change, negative shifting of CB and charge transfer channel formation between heptazine units. Nanosheets synthesized by *in situ* method benefit in the water splitting due to the presence of reactive sites over the large surface area [127–129]. Plasmonic Ag nanoparticle along with B in g-C_3N_4 increased the activity by 4.2 times which is mainly due to reduction in recombination by charge trap site generation, increased electron mobility and band gap decrement resulting in higher visible region absorption. Therefore, synergetic effect of Ag-B significantly improved the photocatalytic activity compared to mono-doped system [79]. Porous structure, enlarged surface area and interstitial doping of electron-rich Na and P together provided improved charge migration, transfer and mobility due to electron affinity of alkali metal as seen by decreased average lifetime to 5.42 ns indicating emergence of non-radiative charge transfer pathways resulting in efficient photoactivity [130]. In another study where P was co-doped with alkaline earth metal Ba, in g-C_3N_4 microtubes it exhibits hydrogen evolution up to 12 µmol/h under visible region greater than 420 nm which conforms its 13.2-fold increment to pristine. This increment is attributing to delocalized density of states distribution over HOMO–LUMO and larger surface area for higher photogenerated charge carriers [131]. For the enhancement of photocatalytic activity of tri-s-triazine based g-C_3N_4, an effective approach of metal (Co) loading and non-metal (B) doping has been considered. On the basis of increment in charge

Photocatalyst 15

carrier generation, reduction in recombination rate and charge redistribution over the surface of g-C_3N_4, two different interaction scenarios (short range and long range) for Co and B were studied and explained. Partial charge compensation along with its conversion into n-type semiconductor with Co-loading and the formation of Co-3d intermediate states while B-doping in g-C_3N_4 formed p-type semiconductor. In comparison to g-C_3N_4, all functionalized films have higher visible light absorbance and lower overpotential values for OER, whereas Co–C_3N_4 and co-doped Co–B–C_3N_4 have lower overpotential values for HER [94]. CO_2 reduction utilizing B, K co-doping along with N vacancies has seen photocatalytic production of 3.16 µmol/g CO and 5.93 µmol/g CH_4, respectively. This high production was achieved using H_2O and CO_2 as feedstocks which are 527% and 161% of CO and CH_4 produced by pristine g-C_3N_4 without any organic hole scavenger preferring co-doping. This is caused due to the synergistic effect of co-doping over mono-doping which lacks electron donation sites and also the co-doping facilitates CO_2 adsorption over surface [132]. Doping of anion and cation along with the co-doping affects the band edges of the pristine semiconductor which are compiled in wa and b, and it is clearly seen that VB and CB edges are shifted significantly toward the redox potentials.

1.3.3 Semiconductor Heterostructure and Metallic Co-catalyst

Considering the points included by the United Nations in sustainable goals for affordable and clean energy accompanied by air purification utilizing natural resources and looking at the current requirement of fuel usage at the world level for transport, industry, power sector and medical applicability, Government of India has also launched the National Hydrogen Mission inviting various industrialists and academicians to collaborate over the design and development of such techniques and for their suitable day-to-day life use. Undergoing the literature, we found that photocatalytic energy conversion displays an attractive pathway to convert earth-abundant molecules (e.g. H_2O, CO_2, or N_2) into fuels and high-value products (e.g. H_2, hydrocarbons, oxygenates, or NH_3). The driving force obtained from electron–hole pairs generated in semiconducting materials upon light irradiation is the basis of photocatalysis. Meanwhile, its performance can also be improved by the formation of a cascading junction (Z-scheme) by alignment of same/different material films, hence forming bilayer or heterostructure aimed at enhancing visible light absorption and promoting photogenerated electron–hole separation [133,134]. Formation of heterostructure with narrow band gaps and suitable band-edge positions for utilization of the merit of each component is most viable. Among all three types including Schottky Junction, type II heterojunction and Z-scheme heterojunction as presented in Figure 1.6, type II heterojunction not only provides fast charge transfer route but also suppresses the rapid recombination of electron–hole pairs increasing the photo-conversion efficiency. Due to promotion of charge separation in type II heterojunction by sacrificing charge carriers, a high thermodynamic requirement of CO_2RR makes it unfavourable for feasibility of the reaction mechanism. Natural photosynthesis uses an efficient mode for the transport of photogenerated electrons through a Z-scheme mode [134]. Although type II heterostructure and Z-scheme have similar band structures where one semiconductor suitable for reduction reaction has CB position whereas other

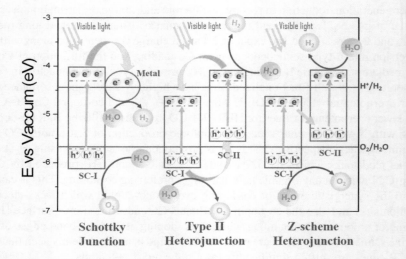

FIGURE 1.6 Charge transfer mechanism in different types of heterojunctions.

semiconductors with low VB position display feasibility for oxidation reaction. When light irradiates on the combination, electrons are excited from VBs to the CBs of both materials and recombination of charge carriers takes place across the interface, thus following a special "Z" shape transport pathway which preserves the pair by strong redox capability, and also the Z-scheme enables wide light response range [135,136].

Numerous works done on various forms of heterojunction have been found which need another chapter to completely review them on the present theme. Thus, the most recent and most efficient heterostructure that plays a critical role in the future of photocatalysis has been considered in this chapter that includes recently developed, synthesized and most popular family of 2D-layered compounds known as MXenes that are both conductors and semiconductors depending on the constituent transition metal and surface functional group. MXenes-based photocatalyst exhibit accelerated charge separation and suppression of carrier recombination [137] as their Fermi level is compatible with that of commonly used wide band gap photocatalysts like ZnO and TiO_2 for the formation of efficient heterojunction. Fermi alignment encourages electron migration to MXene surfaces henceforth formation of Schottky barrier at the interface which prevents electron back-flow on the formation of heterojunction. Therefore, MXenes should be a prime choice as co-catalysts in their metallic form as well as in semiconductor form in formation of Z-scheme for interlayer CB to VB charge transfer and increasing carrier lifetimes for longer reactivity over the surface. It is found that MXenes, especially Sc based such as $Sc_2C(OH)_2$, are speculated to display high HER photocatalytic activity, attributing to an indirect band gap of around 2.0 eV and CBM above HER theoretical potential making them suitable for Z-scheme semiconductor for the formation of heterostructure with $g-C_3N_4$ [138]. Ti_3C_2 nanoparticles are highly efficient co-catalyst in H_2 production (14,342 μmol/h g) and have a high quantum efficiency of 40.1% at 420 nm. In addition, they serve efficiently on ZnS or $Zn_xCd_{1-x}S$ [139].

First-principles-based investigation utilizing hybrid functional has been performed for the photocatalytic properties of 2D g-C_3N_4/Ti_2CO_2 van der Waals heterostructure. It falls under the category of type II heterostructure showing band alignment staggering and formation of built-in interfacial electric field due to enhanced charge transfer to Ti_2CO_2 from g-C_3N_4. High light-harvesting capability, suppression of charge carrier recombination and increased catalytic activity of g-C_3N_4/Ti_2CO_2 towards both HER and OER were demonstrated and reported by reduced overpotential values attributing to the spatially distinguished VB and CB edges, promoting the possibility of direct Z-scheme photocatalytic mechanism [140]. The work done by Liu et al. [141] by fabricating g-C_3N_4/Ti_3C_2 composite resulted in 2.75 times greater photocurrent and enhanced photocatalytic capability of ciprofloxacin degradation as compared to g-C_3N_4. By varying the mixed ratio of g-C_3N_4 with Ti_3C_2 and modifying the surface termination groups on the surface of Ti_3C_2, it was investigated as a co-catalyst in HER. The best HER results were obtained by annealing in the air in a muffle furnace with a high H_2 production of 88 μmol/h g for g-C_3N_4/Ti_3C_2 in a ratio of 190%. Ling et al. [142] have investigated the HER performance step by step for fully O-terminated MXenes, including 7 bi-metal carbides and 10 mono-metal carbides. In another study, lamellar g-C_3N_4 on Ti_3C_2 surface synthesized by Li et al. [143] have reported the formation of Schottky barrier for facilitating electron immigration attributing to excellent conductivity of MXene at the interface and gives 6 times higher hydrogen production than pure pristine material [143]. He et al. [144] synthesized TiO_2/C_3N_4 (2D/2D) core-shell van der Waals heterojunction and deposited 0D MXene forming S-scheme charge transfer pathway between 2D structure and trapping electrons from C_3N_4. This study provides multi-junction interface tuning for controlled charge migration and regulating enhanced CO_2RR [144]. Work done on the alkalinized Ti_3C_2 with g-C_3N_4 decoration attributed to superior electrical conductivity and large Fermi level difference for photo-induced carrier separation for 5.9-fold photocatalytic CO_2RR than pristine g-C_3N_4 and suggests MXene as low-cost and noble metal-free co-catalyst for photocatalysis [145]. Bai et al. [146] synthesized a Z-scheme g-C_3N_4/$Bi_4O_5I_2$ photocatalyst based on good photocatalytic nature of bismuth oxyhalides for CO_2 reduction with I^{-3}/I^- as redox mediators. In the absence of co-catalyst and sacrificial regent, the optimized g-C_3N_4/$Bi_4O_5I_2$ system gave a 45.6 μmol/h g production rate of CO quite high as compared to pristine samples [146]. There are other heterostructures explored both experimentally and theoretically based on g-C_3N_4/ZnO for high activity of CO, g-C_3N_4/α−Fe_2O_3 [147], also the high activity of methane and methanol by g-C_3N_4/Sn_2S_3-DETA [148]. Various co-doping strategies in heterojunctions have also been brought forward theoretically and experimentally for enhanced photocatalytic activity promoting the interlayer charge transfer, utilizing interfacial coupling Z-scheme charge migration pathway to decrease the recombination along with formation of charge-trapping sites in the forbidden region in the study of (Cr, B) co-doped g-C_3N_4/$BiVO_4$ Z-scheme heterostructure by Wang et al. [149]. Zhao et al. [150] have investigated the photocatalytic mechanism of (F, Ti) co-doped heptazine/triazine-based g-C_3N_4 heterostructure using hybrid DFT. They have found narrowing of calculated band gap which is advantageous for absorbing visible spectrum of irradiated light resulting in an obvious red shift of optical absorption edge [150], while in another study, spin-polarized DFT+U method is used to check the photocatalytic performance of (Cu, N) co-doped TiO_2/ g-C_3N_4 heterostructure resulting in reduction of band gap as compared to TiO_2(1 0 1) surface and induction of impurity states of N and Cu appearing in the band gap region of TiO_2/g-C_3N_4 leading to reduction of photon excitation energy [151].

1.4 ROLE OF CHARGE TRANSFER IN ENHANCING PHOTOCATALYTIC ACTIVITY

Charge transfer plays a very important role in understanding the photocatalytic reaction. Apart from the separation of electron–hole pair, transfer of electron from VB to CB, the distribution of charges over the surface and formation of active site for redox reaction is also highly dependent on charge transfer mechanism. Various attempts have been made to explain the formation of charge transfer channel in g-C_3N_4 after modification by doping, loading and heterostructure formation. It is seen that the overpotential of HER and OER depends on the bonding between intermediates (OH, O, OOH, H_3O and H) and reaction site. More electronegative atom on the surface of the photocatalyst participates more actively in the redox reaction and gives low value of overpotential as compared to the other atoms of the photocatalyst surface [152–155]. In the case of P- and S-doped g-C_3N_4, a charge transfer channel via the N–S–N–C–N–P pathway is suggested to enhance the photocatalytic activity [127]. In the study of TM doping in g-C_3N_4, transition metal atom is considered as regulator to adjust the electronegativity of the active site and hence the reduction in overpotential is reported [127]. The charge transfer path by the Co and B doping in g-C_3N_4 as mentioned in Figure 1.7 and Co-B together through a charge transfer channel as marked by the red arrow from neighbouring N to C and then the most electronegative

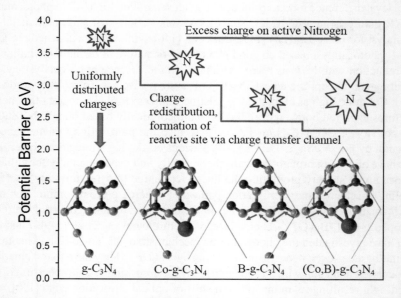

FIGURE 1.7 Role of electronegativity and charge transfer channel in pristine, B-doped, Co-loaded and (Co,B)– g-C_3N_4 monolayer for decrement of a potential barrier in OER [94].

N [94] make it most suitable for redox reactions with reduction in potential barrier to overcome the prohibitive overpotential [152] as compared to pristine g-C_3N_4.

1.5 CONCLUSION

Carbon nitride is considered as the most popular, stable, low-cost, metal-free photocatalyst under visible light, but due to its high charge carrier recombination, various successful attempts have been made to enhance the photocatalytic efficiency by anion doping, cation doping, anion–cation co-doping, formation of heterostructure, or loading of co-catalyst to form Schottky barrier and heterojunctions as reviewed in this chapter. Effects of various functionalization methods on band edges, optical properties, overall charge distribution on the surface of g-C_3N_4 and effect on overpotential have also been discussed in terms of HER, OER and CO_2RR. For better understanding of optical properties, studies based on excitonic effects are highly required in this field. For better understanding of reaction mechanism, nudged elastic band method is also highly required for functionalized g-C_3N_4 especially to determine the role of charge carrier transfer on adsorption of an intermediate molecule and to have some elaborative insights to reduce the barrier height of rate-determining reaction.

REFERENCES

1. Mao, S. S., & Shen, S. (2013). Catalysing artificial photosynthesis. *Nature Photonics*, 7(12), 944–946.
2. Cox, N., Pantazis, D. A., Neese, F., & Lubitz, W. (2015). Artificial photosynthesis: Understanding water splitting in nature. *Interface Focus*, 5(3), 20150009–20150018.
3. Whang, D. R., & Apaydin, D. H. (2018). Artificial photosynthesis: Learning from nature. *ChemPhotoChem*, 2(3), 148–160.
4. Xu, Y., Li, A., Yao, T., Ma, C., Zhang, X., Shah, J.H., & Han, H. (2017). Strategies for efficient charge separation and transfer in artificial photosynthesis of solar fuels. *Chemistry Sustainability Energy Materials*, 10(22), 4277–4305.
5. Handoko, A. D., Li, K., & Tang, J. (2013). Recent progress in artificial photosynthesis: CO_2 photoreduction to valuable chemicals in a heterogeneous system. *Current Opinion in Chemical Engineering*, 2(2), 200–206.
6. AlOtaibi, B., Fan, S., Wang, D., Ye, J., & Mi, Z. (2015). Wafer-level artificial photosynthesis for CO_2 reduction into CH_4 and CO using GaN nanowires. *ACS Catalysis*, 5(9), 5342–5348.
7. Liu, Q., Yuan, J., Gan, Z., Liu, C., Li, J., Liang, Y., & Chen, R. (2020). Photocatalytic N_2 reduction: Uncertainties in the determination of ammonia production. *ACS Sustainable Chemistry Engineering*, 9(1), 560–568.
8. Ren, H., Koshy, P., Chen, W. F., Qi, S., & Sorrell, C. C. (2017). Photocatalytic materials and technologies for air purification. *Journal of Hazardous Materials*, 325, 340–366.
9. Mamaghani, A. H., Haghighat, F., & Lee, C. S. (2017). Photocatalytic oxidation technology for indoor environment air purification: The state-of-the-art. *Applied Catalysis B: Environmental*, 203, 247–269.
10. Shen, W., Zhang, C., Li, Q., Zhang, W., Cao, L., & Ye, J. (2015). Preparation of titanium dioxide nano particle modified photocatalytic self-cleaning concrete. *Journal of Cleaner Production*, 87, 762–765.
11. Hüsken, G., Hunger, M., & Brouwers, H. J. H. (2009). Experimental study of photocatalytic concrete products for air purification. *Building and Environment*, 44(12), 2463–2474.

12. Auvinen, J., & Wirtanen, L. (2008). The influence of photocatalytic interior paints on indoor air quality. *Atmospheric Environment*, 42(18), 4101–4112.
13. Hochmannova, L., & Vytrasova, J. (2010). Photocatalytic and antimicrobial effects of interior paints. *Progress in Organic Coatings*, 67(1), 1–5.
14. Idris, A. M., Liu, T., Shah, J. H., Zhang, X., Ma, C., Malik, A. S., Jin, A., Rasheed, S., Sun, Y., Li, C., & Han, H. (2020). A novel double perovskite oxide semiconductor Sr_2CoWO_6 as bifunctional photocatalyst for photocatalytic oxygen and hydrogen evolution reactions from water under visible light irradiation. *Solar RRL*, 4(3), 1900456–1900466.
15. Idris, A. M., Liu, T., Hussain Shah, J., Han, H., & Li, C. (2020). Sr_2CoTaO_6 double perovskite oxide as a novel visible-light-absorbing bifunctional photocatalyst for photocatalytic oxygen and hydrogen evolution reactions. *ACS Sustainable Chemistry Engineering*, 8(37), 14190–14197.
16. Abanades, S. (2019). Metal oxides applied to thermochemical water-splitting for hydrogen production using concentrated solar energy. *Chemical Engineering*, 3(3), 63–90.
17. Chen, Y., Feng, X., Liu, Y., Guan, X., Burda, C., & Guo, L. (2020). Metal oxidebased tandem cells for self-biased photoelectrochemical water splitting. *ACS Energy Letters*, 5(3), 844–866.
18. Geiss, O., Cacho, C., Barrero-Moreno, J., & Kotzias, D. (2012). Photocatalytic degradation of organic paint constituents-formation of carbonyls. *Building and Environment*, 48, 107–112.
19. Kandavelu, V., Kastien, H., & Thampi, K. R. (2004). Photocatalytic degradation of isothiazolin-3-ones in water and emulsion paints containing nanocrystalline TiO_2 and ZnO catalysts. *Applied Catalysis B: Environmental*, 48(2), 101–111.
20. Barka, N., Qourzal, S., Assabbane, A., Nounah, A., & Ait-Ichou, Y. (2010). Photocatalytic degradation of an azo reactive dye, Reactive Yellow 84, in water using an industrial titanium dioxide coated media. *Arabian Journal of Chemistry*, 3(4), 279–283.
21. Jain, R., & Shrivastava, M. (2008). Photocatalytic removal of hazardous dye cyanosine from industrial waste using titanium dioxide. *Journal of Hazardous Materials*, 152(1), 216–220.
22. Monga, D., & Basu, S. (2019). Enhanced photocatalytic degradation of industrial dye by $g-C_3N_4/TiO_2$ nanocomposite: Role of shape of TiO_2. *Advanced Powder Technology*, 30(5), 1089–1098.
23. Mishra, A., Mehta, A., Kainth, S., & Basu, S. (2018). Effect of $g-C_3N_4$ loading on TiO_2/Bentonite nanocomposites for efficient heterogeneous photocatalytic degradation of industrial dye under visible light. *Journal of Alloys and Compounds*, 764, 406–415.
24. Akhavan, O., & Ghaderi, E. (2010). Self-accumulated Ag nanoparticles on mesoporous TiO_2 thin film with high bactericidal activities. *Surface and Coatings Technology*, 204(21–22), 3676–3683.
25. Lee, H. S., Im, S. J., Kim, J. H., Kim, H. J., Kim, J. P., & Min, B. R. (2008). Polyamide thin-film nanofiltration membranes containing TiO_2 nanoparticles. *Desalination*, 219(1–3), 48–56.
26. Pal, B., & Sharon, M. (2000). Photodegradation of polyaromatic hydrocarbons over thin film of TiO_2 nanoparticles; a study of intermediate photoproducts. *Journal of Molecular Catalysis A: Chemical*, 160(2), 453–460.
27. Wang, G., Ling, Y., Wang, H., Yang, X., Wang, C., Zhang, J. Z., & Li, Y. (2012). Hydrogen-treated WO_3 nanoflakes show enhanced photostability. *Energy Environmental Science*, 5(3), 6180–6187.
28. Kim, J., Lee, C. W., & Choi, W. (2010). Platinized WO_3 as an environmental photocatalyst that generates OH radicals under visible light. *Environmental Science Technology*, 44(17), 6849–6854.
29. Kwon, Y. T., Song, K. Y., Lee, W. I., Choi, G. J., & Do, Y. R. (2000). Photocatalytic behavior of WO_3-loaded TiO_2 in an oxidation reaction. *Journal of Catalysis*, 191(1), 192–199.

30. Tian, C., Zhang, Q., Wu, A., Jiang, M., Liang, Z., Jiang, B., & Fu, H. (2012). Cost effective large-scale synthesis of ZnO photocatalyst with excellent performance for dye photodegradation. *Chemical Communications*, 48(23), 2858–2860.
31. Yang, L. Y., Dong, S. Y., Sun, J. H., Feng, J. L., Wu, Q. H., & Sun, S. P. (2010). Microwave-assisted preparation, characterization and photocatalytic properties of a dumbbell-shaped ZnO photocatalyst. *Journal of Hazardous Materials*, 179(1–3), 438–443.
32. Chen, X., Wu, Z., Liu, D., & Gao, Z. (2017). Preparation of ZnO photocatalyst for the efficient and rapid photocatalytic degradation of azo dyes. *Nanoscale Research Letters*, 12(1), 1–10.
33. Cheng, L., Xiang, Q., Liao, Y., & Zhang, H. (2018). CdS-based photocatalysts. *Energy Environmental Science*, 11(6), 1362–1391.
34. Fard, N. E., Fazaeli, R., & Ghiasi, R. (2016). Band gap energies and photocatalytic properties of CdS and Ag/CdS nanoparticles for Azo dye degradation. *Chemical Engineering Technology*, 39(1), 149–157.
35. Shah, E., & Soni, H. P. (2013). Inducing chirality on ZnS nanoparticles for asymmetric aldol condensation reactions. *RSC Advances*, 3(38), 17453–17461.
36. Chen, D., Huang, F., Ren, G., Li, D., Zheng, M., Wang, Y., & Lin, Z. (2010). ZnS nano-architectures: Photocatalysis, deactivation and regeneration. *Nanoscale*, 2(10), 2062–2064.
37. Lashgari, H., Boochani, A., Shekaari, A., Solaymani, S., Sartipi, E., & Mendi, R.T. (2016). Electronic and optical properties of 2D graphene-like ZnS: DFT calculations. *Applied Surface Science*, 369, 76–81.
38. Riaz, S., Ashraf, M., Hussain, T., Hussain, M. T., & Younus, A. (2019). Fabrication of robust multifaceted textiles by application of functionalized TiO_2 nanoparticles. *Colloids and Surfaces A: Physicochemical and Engineering Aspects*, 581, 123799–123811.
39. Yuzer, B., Guida, M., Ciner, F., Aktan, B., Aydin, M. I., Meric, S., & Selcuk, H. (2016). A multifaceted aggregation and toxicity assessment study of sol–gel-based TiO_2 nanoparticles during textile wastewater treatment. *Desalination and Water Treatment*, 57(11), 4966–4973.
40. Zhan, X., Luo, Y., Wang, Z., Xiang, Y., Peng, Z., Han, Y., Zhang, H., Chen, R., Zhou, Q., Peng, H., & Huang, H. (2022). Formation of multifaceted nano-groove structure on rutile TiO_2 photoanode for efficient electron-hole separation and water splitting. *Journal of Energy Chemistry*, 65, 19–25.
41. Ma, Y., Wong, C. P., Zeng, X. T., Yu, T., Zhu, Y., & Shen, Z. X. (2009). Pulsed laser deposition of ZnO honeycomb structures on metal catalyst prepatterned Si substrates. *Journal of Physics D: Applied Physics*, 42(6), 065417.
42. Catlow, C. R. A., French, S. A., Sokol, A. A., Al-Sunaidi, A. A., & Woodley, S. M. (2008). Zinc oxide: A case study in contemporary computational solid state chemistry. *Journal of Computational Chemistry*, 29(13), 2234–2249.
43. Liebig, J. V. (1834). About some nitrogen compounds. *Annales Pharmaceutiques Francaises*, 10(10), 10.
44. Liu, A. Y., & Cohen, M. L. (1989). Prediction of new low compressibility solids. *Science*, 245(4920), 841–842.
45. Liu, A. Y., & Wentzcovitch, R. M. (1994). Stability of carbon nitride solids. *Physical Review B*, 50(14), 10362–10365.
46. Teter, D. M., & Hemley, R. J. (1996). Low-compressibility carbon nitrides. *Science*, 271(5245), 53–55.
47. Alves, I., Demazeau, G., Tanguy, B., & Weill, F. (1999). On a new model of the graphitic form of C_3N_4. *Solid State Communications*, 109(11), 697–701.
48. Matar, S. F., & Mattesini, M. (2001). Ab initio search of carbon nitrides, isoelectronic with diamond, likely to lead to new ultra hard materials. *Comptes Rendus de l'Acad'emie des Sciences-Series IIC-Chemistry*, 4(4), 255–272.

49. Komatsu, T. (2001). Prototype carbon nitrides similar to the symmetric triangular form of melon. *Journal of Materials Chemistry*, 11(3), 802–803.
50. Xu, Y., & Gao, S. P. (2012). Band gap of C_3N_4 in the GW approximation. *International Journal of Hydrogen Energy*, 37(15), 11072–11080.
51. Kroke, E., Schwarz, M., Horath-Bordon, E., Kroll, P., Noll, B., & Norman, A. D. (2002). Tri-s-triazine derivatives. Part I. From trichloro-tri-s-triazine to graphitic C_3N_4 structures. *New Journal of Chemistry*, 26(5), 508–512.
52. Nabok, D., Puschnig, P., & Ambrosch-Draxl, C., (2008). Cohesive and surface energies of π-conjugated organic molecular crystals: A first-principles study. *Physical Review B*, 77(24), 245316–245319.
53. Inada, Y., Amaya, T., Shimizu, Y., Saeki, A., Otsuka, T., Tsuji, R., Seki, S., & Hirao, T. (2013). Nitrogen-doped graphitic carbon synthesized by laser annealing of Sumanenemonoone Imine as a bowl-shaped π-conjugated molecule. *Chemistry–An Asian Journal*, 8(11), 2569–2574.
54. Xia, P., Cheng, B., Jiang, J., & Tang, H. (2019). Localized π-conjugated structure and EPR investigation of g-C_3N_4 photocatalyst. *Applied Surface Science*, 487, 335–342.
55. Raza, W., Bahnemann, D., & Muneer, M. (2017). Efficient visible light driven, mesoporous graphitic carbon nitrite based hybrid nanocomposite: With superior photocatalytic activity for degradation of organic pollutant in aqueous phase. *Journal of Photochemistry and Photobiology A: Chemistry*, 342, 102–115.
56. Chen, F., Yang, Q., Wang, Y., Zhao, J., Wang, D., Li, X., Guo, Z., Wang, H., Deng, Y., Niu, C., & Zeng, G. (2017). Novel ternary heterojunction photococatalyst of Ag nanoparticles and g-C_3N_4 nanosheets co-modified $BiVO_4$ for wider spectrum visible-light photocatalytic degradation of refractory pollutant. *Applied Catalysis B: Environmental*, 205, 133–147.
57. Tuna, O., & Simsek, E. B. (2020). Synergic contribution of intercalation and electronic modification of g-C_3N_4 for an efficient visible light-driven catalyst for tetracycline degradation. *Journal of Environmental Chemical Engineering*, 8(5), 104445–104455.
58. Jiang, L. L., Wang, Z. K., Li, M., Zhang, C. C., Ye, Q. Q., Hu, K. H., Lu, D. Z., Fang, P. F., & Liao, L. S. (2018). Passivated perovskite crystallization via g-C_3N_4 for high-performance solar cells. *Advanced Functional Materials*, 28(7), 1705875.
59. Zhang, X., Wang, H., Wang, H., Zhang, Q., Xie, J., Tian, Y., Wang, J., & Xie, Y. (2014). Single-layered graphitic-C_3N_4 quantum dots for two-photon fluorescence imaging of cellular nucleus. *Advanced Materials*, 26(26), 4438–4443.
60. Sun, B., Lu, N., Su, Y., Yu, H., Meng, X., & Gao, Z. (2017). Decoration of TiO_2 nanotube arrays by graphitic-C_3N_4 quantum dots with improved photoelectrocatalytic performance. *Applied Surface Science*, 394, 479–487.
61. Su, Y., Sun, B., Chen, S., Yu, H., & Liu, J. (2017). Fabrication of graphitic-C_3N_4 quantum dots coated silicon nanowire array as a photoelectrode for vigorous degradation of 4-chlorophenol. *RSC Advances*, 7(24), 14832–14836.
62. Zhao, Y., Zhao, F., Wang, X., Xu, C., Zhang, Z., Shi, G., & Qu, L. (2014). Graphitic carbon nitride nanoribbons: Graphene-assisted formation and synergic function for highly efficient hydrogen evolution. *Angewandte Chemie International Edition*, 53(50), 13934–13939.
63. Tahir, M., Cao, C., Mahmood, N., Butt, F. K., Mahmood, A., Idrees, F., Hussain, S., Tanveer, M., Ali, Z., & Aslam, I. (2014). Multifunctional g-C_3N_4 nanofibers: A template-free fabrication and enhanced optical, electrochemical, and photocatalyst properties. *ACS Applied Materials Interfaces*, 6(2), 1258–1265.
64. Wang, S., Li, C., Wang, T., Zhang, P., Li, A., & Gong, J. (2014). Controllable synthesis of nanotube-type graphitic C_3N_4 and their visible-light photocatalytic and fluorescent properties. *Journal of Materials Chemistry A*, 2(9), 2885–2890.

65. Wang, Y., Wang, H., Chen, F., Cao, F., Zhao, X., Meng, S., & Cui, Y. (2017). Facile synthesis of oxygen doped carbon nitride hollow microsphere for photocatalysis. *Applied Catalysis B: Environmental*, 206, 417–425.
66. Sun, J., Zhang, J., Zhang, M., Antonietti, M., Fu, X., & Wang, X. (2012). Bioinspired hollow semiconductor nanospheres as photosynthetic nanoparticles. *Nature Communications*, 3(1), 1–7.
67. Feng, L. L., Zou, Y., Li, C., Gao, S., Zhou, L. J., Sun, Q., Fan, M., Wang, H., Wang, D., Li, G. D., & Zou, X. (2014). Nanoporous sulfur-doped graphitic carbon nitride microrods: A durable catalyst for visible-light-driven H_2 evolution. *International Journal of Hydrogen Energy*, 39(28), 15373–15379.
68. You, Z., Su, Y., Yu, Y., Wang, H., Qin, T., Zhang, F., Shen, Q., & Yang, H. (2017). Preparation of g-C_3N_4 nanorod/$InVO_4$ hollow sphere composite with enhanced visible-light photocatalytic activities. *Applied Catalysis B: Environmental*, 213, 127–135.
69. Tahir, B., Tahir, M., & Amin, N. A. S. (2017). Photo-induced CO_2 reduction by CH_4/H_2O to fuels over Cu-modified g-C_3N_4 nanorods under simulated solar energy. *Applied Surface Science*, 419, 875–885.
70. Tahir, M., Cao, C., Butt, F. K., Butt, S., Idrees, F., Ali, Z., Aslam, I., Tanveer, M., Mahmood, A., & Mahmood, N. (2014). Large scale production of novel g-C_3N_4 micro strings with high surface area and versatile photodegradation ability. *CrystEngComm*, 16(9), 1825–1830.
71. Narkbuakaew, T., & Sujaridworakun, P. (2020). Synthesis of Tri-S-triazine based g-C_3N_4 photocatalyst for cationic rhodamine B degradation under visible light. *Topics in Catalysis*, 63(11), 1086–1096.
72. Ravichandran, D., Akilan, R., Vinnarasi, S., Shankar, R., & Manickam, S. (2021). Tuning the reactivity of tri-s-triazine, trinitro-tri-s-triazine and ternary tri-s-triazine graphitic C_3N_4 quantum dots through H-functionalized and B-doped complexes: A density functional study. *Chemosphere*, 272, 129901–129910.
73. Zhu, B., Zhang, J., Jiang, C., Cheng, B., & Yu, J. (2017). First principle investigation of halogen-doped monolayer g-C_3N_4 photocatalyst. *Applied Catalysis B: Environmental*, 207, 27–34.
74. Devthade, V., Kulhari, D., & Umare, S. S. (2018). Role of precursors on photocatalytic behavior of graphitic carbon nitride. *Materials Today: Proceedings*, 5(3), 9203–9210.
75. Thorat, N., Borade, S., Varma, R., Yadav, A., Gupta, S., Fernandes, R., Sarawade, P., Bhanage, B. M., & Patel, N. (2021). High surface area Nanoflakes of P– g-C_3N_4 photocatalyst loaded with Ag nanoparticle with intraplanar and interplanar charge separation for environmental remediation. *Journal of Photochemistry and Photobiology A: Chemistry*, 408, 113098–113109.
76. Varma, R., Chaurasia, S., Patel, N., & Bhanage, B. M., 2020. Interplay of adsorption, photo-absorption, electronic structure and charge carrier dynamics on visible light driven photocatalytic activity of Bi_2MoO_6/rGO (0D/2D) heterojunction. *Journal of Environmental Chemical Engineering*, 8(6), 104551.
77. Gupta, S., Patel, M.K., Miotello, A., & Patel, N. (2020). Metal boride-based catalysts for electrochemical water-splitting: A review. *Advanced Functional Materials*, 30(1), 1906481–1906508.
78. Roselin, L. S., Patel, N., & Khayyat, S. A. (2019). Codoped g-C_3N_4 nanosheet for degradation of organic pollutants from oily wastewater. *Applied Surface Science*, 494, 952–958.
79. Thorat, N., Yadav, A., Yadav, M., Gupta, S., Varma, R., Pillai, S., Fernandes, R., Patel, M., & Patel, N. (2019). Ag loaded B-doped-g-C_3N_4 nanosheet with efficient properties for photocatalysis. *Journal of Environmental Management*, 247, 57–66.

80. Jaiswal, R., Patel, N., Dashora, A., Fernandes, R., Yadav, M., Edla, R., Varma, R. S., Kothari, D. C., Ahuja, B. L., & Miotello, A. (2016). Efficient Co-B-codoped TiO_2 photocatalyst for degradation of organic water pollutant under visible light. *Applied Catalysis B: Environmental*, 183, 242–253.
81. Patel, N., Dashora, A., Jaiswal, R., Fernandes, R., Yadav, M., Kothari, D. C., Ahuja, B. L., & Miotello, A. (2015). Experimental and theoretical investigations on the activity and stability of substitutional and interstitial boron in TiO_2 photocatalyst. *The Journal of Physical Chemistry C*, 119(32), 18581–18590.
82. Zhao, Z., & Liu, Q. (2007). Mechanism of higher photocatalytic activity of anatase TiO_2 doped with nitrogen under visible-light irradiation from density functional theory calculation. *Journal of Physics D: Applied Physics*, 41(2), 025105.
83. Yang, J., Wang, D., Zhou, X., & Li, C. (2013). A theoretical study on the mechanism of photocatalytic oxygen evolution on $BiVO_4$ in aqueous solution. *Chemistry*, 19(4), 1320–1326.
84. Fu, C.F., Wu, X., & Yang, J. (2018). Material design for photocatalytic water splitting from a theoretical perspective. *Advanced Materials*, 30(48), 1802106–1802116.
85. Pang, R., Yu, L. J., Wu, D. Y., Mao, B. W., & Tian, Z. Q. (2013). Surface electron– hydronium ion-pair bound to silver and gold cathodes: A density functional theoretical study of photocatalytic hydrogen evolution reaction. *Electrochimica Acta*, 101, 272–278.
86. Yu, K. M. K., Curcic, I., Gabriel, J., & Tsang, S. C. E. (2008). Recent advances in CO_2 capture and utilization. *Chemistry Sustainability Energy Materials*, 1(11), 893–899.
87. Takanabe, K. (2017). Photocatalytic water splitting: Quantitative approaches toward photocatalyst by design. *ACS Catalysis*, 7(11), 8006–8022.
88. Kamat, P. V. (2012). Manipulation of charge transfer across semiconductor interface. A criterion that cannot be ignored in photocatalyst design. *The Journal of Physical Chemistry Letters*, 3(5), 663–672.
89. Acar, C., Dincer, I., & Zamfirescu, C. (2014). A review on selected heterogeneous photocatalysts for hydrogen production. *International Journal of Energy Research*, 38(15), 1903–1920.
90. Ding, K., Chen, B., Li, Y., Zhang, Y., & Chen, Z. (2014). Comparative density functional theory study on the electronic and optical properties of $BiMO_4$ (M= V, Nb, Ta). *Journal of Materials Chemistry A*, 2(22), 8294–8303.
91. Norskov, J. K., Bligaard, T., Logadottir, A., Kitchin, J. R., Chen, J. G., Pandelov, S., & Stimming, U. (2005). Trends in the exchange current for hydrogen evolution. *Journal of the Electrochemical Society*, 152(3), J23.
92. Gao, G., Jiao, Y., Ma, F., Jiao, Y., Waclawik, E., & Du, A. (2015). Metal-free graphitic carbon nitride as mechano-catalyst for hydrogen evolution reaction. *Journal of Catalysis*, 332, 149–155.
93. Rossmeisl, J., Logadottir, A., & Norskov, J. K. (2005). Electrolysis of water on (oxidized) metal surfaces. *Chemical Physics*, 319(1–3), 178–184.
94. Bhagat, B.R., & Dashora, A. (2021). Understanding the synergistic effect of Co loading and B-doping in $g-C_3N_4$ for enhanced photocatalytic activity for overall solar water splitting. *Carbon*, 178, 666–677.
95. Lingampalli, S. R., Ayyub, M. M., & Rao, C. N. R. (2017). Recent progress in the photocatalytic reduction of carbon dioxide. *ACS Omega*, 2(6), 2740–2748.
96. Nørskov, J. K., Rossmeisl, J., Logadottir, A., Lindqvist, L. R. K. J., Kitchin, J. R., Bligaard, T., & Jonsson, H. (2004). Origin of the overpotential for oxygen reduction at a fuel-cell cathode. *The Journal of Physical Chemistry B*, 108(46), 17886–17892.
97. Hernández-Alonso, M. D., Fresno, F., Suárez, S., & Coronado, J. M. (2009). Development of alternative photocatalysts to TiO_2: Challenges and opportunities. *Energy Environmental Science*, 2(12), 1231–1257.

98. Mills, A., & Le Hunte, S. (1997). An overview of semiconductor photocatalysis. *Journal of Photochemistry and Photobiology A: Chemistry*, 108(1), 1–35.
99. Lu, S., Li, C., Li, H. H., Zhao, Y. F., Gong, Y. Y., Niu, L. Y., Liu, X. J., & Wang, T. (2017). The effects of nonmetal dopants on the electronic, optical and chemical performances of monolayer g-C_3N_4 by first-principles study. *Applied Surface Science*, 392, 966–974.
100. Ling, F., Li, W., & Ye, L. (2019). The synergistic effect of non-metal doping or defect engineering and interface coupling on the photocatalytic property of g-C_3N_4: First-principle investigations. *Applied Surface Science*, 473, 386–392.
101. Wang, Y., Tian, Y., Yan, L., & Su, Z. (2018). DFT study on sulfur-doped g-C_3N_4 nanosheets as a photocatalyst for CO_2 reduction reaction. *The Journal of Physical Chemistry C*, 122(14), 7712–7719.
102. Zhu, Y. P., Ren, T. Z., & Yuan, Z. Y. (2015). Mesoporous phosphorus-doped g-C_3N_4 nanostructured flowers with superior photocatalytic hydrogen evolution performance. *ACS Applied Materials Interfaces*, 7(30), 16850–16856.
103. Arumugam, M., Tahir, M., & Prasertdham, P. (2022). Effect of nonmetals (B, O, P, and S) doped with porous g-C_3N_4 for improved electron transfer towards photocatalytic CO_2 reduction with water into CH_4. *Chemosphere*, 286, 131765–131775.
104. Lu, S., Chen, Z.W., Li, C., Li, H. H., Zhao, Y. F., Gong, Y. Y., Niu, L. Y., Liu, X. J., Wang, T., & Sun, C. Q. (2016). Adjustable electronic performances and redox ability of a g-C_3N_4 monolayer by adsorbing nonmetal solute ions: A first principles study. *Journal of Materials Chemistry A*, 4(38), 14827–14838.
105. Kong, L., Mu, X., Fan, X., Li, R., Zhang, Y., Song, P., Ma, F., & Sun, M. (2018). Site-selected N vacancy of g-C_3N_4 for photocatalysis and physical mechanism. *Applied Materials Today*, 13, 329–338.
106. Cui, J., Liang, S., Wang, X., & Zhang, J. (2015). First principle modelling of oxygen doped monolayer graphitic carbon nitride. *Materials Chemistry and Physics*, 161, 194–200.
107. Zhu, B., Zhang, J., Jiang, C., Cheng, B., & Yu, J. (2017). First principle investigation of halogen-doped monolayer g-C_3N_4 photocatalyst. *Applied Catalysis B: Environmental*, 207, 27–34.
108. Luo, Y., Wang, J., Yu, S., Cao, Y., Ma, K., Pu, Y., Zou, W., Tang, C., Gao, F., & Dong, L. (2018). Nonmetal element doped g-C_3N_4 with enhanced H_2 evolution under visible light irradiation. *Journal of Materials Research*, 33(9), 1268–1278.
109. Zhu, Z., Tang, X., Wang, T., Fan, W., Liu, Z., Li, C., Huo, P., & Yan, Y. (2019). Insight into the effect of co-doped to the photocatalytic performance and electronic structure of g-C_3N_4 by first principle. *Applied Catalysis B: Environmental*, 241, 319–328.
110. Tong, T., He, B., Zhu, B., Cheng, B., & Zhang, L. (2018). First-principle investigation on charge carrier transfer in transition-metal single atoms loaded g-C_3N_4. *Applied Surface Science*, 459, 385–392.
111. Hussain, T., Vovusha, H., Kaewmaraya, T., Karton, A., Amornkitbamrung, V., & Ahuja, R. (2018). Graphitic carbon nitride nano sheets functionalized with selected transition metal dopants: An efficient way to store CO_2. *Nanotechnology*, 29(41), 415502–415600.
112. Bai, K., Cui, Z., Li, E., Ding, Y., Zheng, J., Zheng, Y., & Liu, C. (2020). Adsorption of alkali metals on graphitic carbon nitride: A first-principles study. *Modern Physics Letters B*, 34(32), 2050361–2050372.
113. Li, J., Cui, W., Sun, Y., Chu, Y., Cen, W., & Dong, F. (2017). Directional electron delivery via a vertical channel between g-C_3N_4 layers promotes photocatalytic efficiency. *Journal of Materials Chemistry A*, 5(19), 9358–9364.

114. Xiong, T., Cen, W., Zhang, Y., & Dong, F. (2016). Bridging the g-C_3N_4 interlayers for enhanced photocatalysis. *ACS Catalysis*, 6(4), 2462–2472.
115. Jiang, J., Cao, S., Hu, C., & Chen, C. (2017). A comparison study of alkali metal doped g-C_3N_4 for visible-light photocatalytic hydrogen evolution. *Chinese Journal of Catalysis*, 38(12), 1981–1989.
116. Nguyen, T. T. H., Le, M. C., & Ha, N. N. (2021). Understanding the influence of single metal (Li, Mg, Al, Fe, Ag) doping on the electronic and optical properties of g-C_3N_4: A theoretical study. *Molecular Simulation*, 47(1), 10–17.
117. Zhang, W., Cho, H. Y., Zhang, Z., Yang, W., Kim, K. K., & Zhang, F. (2016). First-principles calculation the electronic structure and the optical properties of Mn decorated g-C_3N_4 for photocatalytic applications. *Journal of the Korean Physical Society*, 69(9), 1445–1449.
118. Zhang, W., Zhang, Z., Kwon, S., Zhang, F., Stephen, B., Kim, K. K., Jung, R., Kwon, S., Chung, K. B., & Yang, W. (2017). Photocatalytic improvement of Mn adsorbed g-C_3N_4. *Applied Catalysis B: Environmental*, 206, 271–281.
119. Hu, S., Ma, L., You, J., Li, F., Fan, Z., Lu, G., Liu, D., & Gui, J. (2014). Enhanced visible light photocatalytic performance of g-C_3N_4 photocatalysts co-doped with iron and phosphorus. *Applied Surface Science*, 311, 164–171.
120. Hu, J., Zhang, P., An, W., Liu, L., Liang, Y., & Cui, W. (2019). In-situ Fe-doped g-C_3N_4 heterogeneous catalyst via photocatalysis-Fenton reaction with enriched photocatalytic performance for removal of complex wastewater. *Applied Catalysis B: Environmental*, 245, 130–142.
121. Tonda, S., Kumar, S., Kandula, S., & Shanker, V. (2014). Fe-doped and-mediated graphitic carbon nitride nanosheets for enhanced photocatalytic performance under natural sunlight. *Journal of Materials Chemistry A*, 2(19), 6772–6780.
122. Li, Z., Kong, C., & Lu, G. (2016). Visible photocatalytic water splitting and photocatalytic two-electron oxygen formation over Cu-and Fe-doped g-C_3N_4. *The Journal of Physical Chemistry C*, 120, 56–63.
123. Zhao, X., Wang, L., & Pei, Y. (2021). Single metal atom catalyst supported on g-C_3N_4 for formic acid dehydrogenation: A combining density functional theory and machine learning study. *The Journal of Physical Chemistry C*, 125, 22513–22521.
124. Yang, C., Zhao, Z., & Liu, Q. (2021). Mechanistic insight into the dispersion behaviour of single platinum atom on monolayer g-C_3N_4 in single-atom catalysts from density functional theory calculations. *Applied Surface Science*, 566, 150697.
125. Li, H., Wu, Y., Li, L., Gong, Y., Niu, L., Liu, X., Wang, T., Sun, C., & Li, C. (2018). Adjustable photocatalytic ability of monolayer g-C_3N_4 utilizing single–metal atom: Density functional theory. *Applied Surface Science*, 457, 735–744.
126. Ma, Z., Cui, Z., Lv, Y., Sa, R., Wu, K., & Li, Q. (2020). Three-in-one: Opened charge-transfer channel, positively shifted oxidation potential, and enhanced visible light response of g-C_3N_4 photocatalyst through K and S Co-doping. *International Journal of Hydrogen Energy*, 45(7), 4534–4544.
127. Hu, C., Hung, W. Z., Wang, M. S., & Lu, P. J. (2018). Phosphorus and sulphur codoped g-C_3N_4 as an efficient metal-free photocatalyst. *Carbon*, 127, 374–383.
128. Chu, Y. C., Lin, T. J., Lin, Y. R., Chiu, W. L., Nguyen, B. S., & Hu, C. (2020). Influence of P, S, O-Doping on g-C_3N_4 for hydrogel formation and photocatalysis: An experimental and theoretical study. *Carbon*, 169, 338–348.
129. Liu, Q., Shen, J., Yu, X., Yang, X., Liu, W., Yang, J., Tang, H., Xu, H., Li, H., Li, Y., & Xu, J. (2019). Unveiling the origin of boosted photocatalytic hydrogen evolution in simultaneously (S, P, O)-Codoped and exfoliated ultrathin g-C_3N_4 nanosheets. *Applied Catalysis B: Environmental*, 248, 84–94.
130. Cao, S., Huang, Q., Zhu, B., & Yu, J. (2017). Trace-level phosphorus and sodium co-doping of g-C_3N_4 for enhanced photocatalytic H_2 production. *Journal of Power Sources*, 351, 151–159.

131. Long, D., Chen, W., Zheng, S., Rao, X., & Zhang, Y. (2020). Barium-and phosphorus-codoped g-C_3N_4 microtubes with efficient photocatalytic H2 evolution under visible light irradiation. *Industrial Engineering Chemistry Research*, 59(10), 4549–4556.
132. Wang, J. C., Hou, Y., Feng, F. D., Wang, W. X., Shi, W., Zhang, W., Li, Y., Lou, H., & Cui, C. X. (2021). A recyclable molten-salt synthesis of B and K co-doped g-C_3N_4 for photocatalysis of overall water vapor splitting. *Applied Surface Science*, 537, 148014–148022.
133. Handoko, A. D., Steinmann, S. N., & Seh, Z. W. (2019). Theory-guided materials design: Two-dimensional MXenes in electro-and photocatalysis. *Nanoscale Horizons*, 4(4), 809–827.
134. Low, J., Jiang, C., Cheng, B., Wageh, S., Al-Ghamdi, A. A., & Yu, J. (2017). A review of direct Z-scheme photocatalysts. *Small Methods*, 1(5), 1700080.
135. Jiang, L., Yuan, X., Zeng, G., Liang, J., Wu, Z., & Wang, H. (2018). Construction of an all-solid-state Z-scheme photocatalyst based on graphite carbon nitride and its enhancement to catalytic activity. *Environmental Science: Nano*, 5(3), 599–615.
136. Zhang, W., Mohamed, A. R., & Ong, W. J. (2020). Z-Scheme photocatalytic systems for carbon dioxide reduction: Where are we now? *Angewandte Chemie International Edition*, 59(51), 22894–22915.
137. Sun, Y., Meng, X., Dall'Agnese, Y., Dall'Agnese, C., Duan, S., Gao, Y., Chen, G., & Wang, X. F. (2019). 2D MXenes as co-catalysts in photocatalysis: Synthetic methods. *Nano-Micro Letters*, 11(1), 1–22.
138. Liu, J. H., Kan, X., Amin, B., Gan, L. Y., & Zhao, Y. (2017). Theoretical exploration of the potential applications of Sc-based MXenes. *Physical Chemistry Chemical Physics*, 19(48), 32253–32261.
139. Ran, J., Gao, G., Li, F. T., Ma, T. Y., Du, A., & Qiao, S. Z. (2017). Ti_3C_2 MXene co-catalyst on metal sulfide photo-absorbers for enhanced visible-light photocatalytic hydrogen production. *Nature Communications*, 8(1), 1–10.
140. Liu, X., Kang, W., Qi, L., Zhao, J., Wang, Y., Wang, L., Wang, W., Fang, L., & Zhou, M. (2021). Two-dimensional g-C_3N_4/Ti_2CO_2 heterostructure as a direct Z-scheme photocatalyst for water splitting: A hybrid density functional theory investigation. *Physica E: Low-Dimensional Systems and Nanostructures*, 134, 114872–114878.
141. Liu, N., Lu, N., Su, Y., Wang, P., & Quan, X. (2019). Fabrication of g-C_3N_4/Ti_3C_2 composite and its visible-light photocatalytic capability for ciprofloxacin degradation. *Separation and Purification Technology*, 211, 782–789.
142. Ling, C., Shi, L., Ouyang, Y., & Wang, J. (2016). Searching for highly active catalysts for hydrogen evolution reaction based on O-terminated MXenes through a simple descriptor. *Chemistry of Materials*, 28(24), 9026–9032.
143. Li, J., Zhao, L., Wang, S., Li, J., Wang, G., & Wang, J. (2020). In situ fabrication of 2D/3D g-C_3N_4/Ti_3C_2 (MXene) heterojunction for efficient visible-light photocatalytic hydrogen evolution. *Applied Surface Science*, 515, 145922–145930.
144. He, F., Zhu, B., Cheng, B., Yu, J., Ho, W., & Macyk, W. (2020). 2D/2D/0D TiO_2/g-C_3N_4/Ti_3C_2 MXene composite S-scheme photocatalyst with enhanced CO_2 reduction activity. *Applied Catalysis B: Environmental*, 272, 119006–119017.
145. Tang, Q., Sun, Z., Deng, S., Wang, H., & Wu, Z. (2020). Decorating g-C_3N_4 with alkalinized Ti_3C_2 MXene for promoted photocatalytic CO_2 reduction performance. *Journal of Colloid and Interface Science*, 564, 406–417.
146. Bai, Y., Ye, L., Wang, L., Shi, X., Wang, P., Bai, W., & Wong, P. K. (2016). g-C_3N_4/$Bi_4O_5I_2$ heterojunction with I_3^-/I^- redox mediator for enhanced photocatalytic CO_2 conversion. *Applied Catalysis B: Environmental*, 194, 98–104.
147. Jiang, Z., Wan, W., Li, H., Yuan, S., Zhao, H., &Wong, P. K. (2018). A hierarchical Z scheme α-Fe_2O_3/g-C_3N_4 hybrid for enhanced photocatalytic CO_2 reduction. *Advanced Materials*, 30(10), 1706108–1706.

148. Huo, Y., Zhang, J., Dai, K., Li, Q., Lv, J., Zhu, G., & Liang, C. (2019). All-solid state artificial Z-scheme porous g-C_3N_4/Sn_2S_3-DETA heterostructure photocatalyst with enhanced performance in photocatalytic CO_2 reduction. *Applied Catalysis B: Environmental*, 241, 528–538.
149. Wang, Q., Lin, Y., Li, P., Ma, M., Maheskumar, V., Jiang, Z., & Zhang, R. (2021). An efficient Z-scheme (Cr, B) codoped g-C_3N_4/$BiVO_4$ photocatalyst for water splitting: A hybrid DFT study. *International Journal of Hydrogen Energy*, 46(1), 247–261.
150. Zhao, Y., Lin, Y., Wang, G., Jiang, Z., Zhang, R., & Zhu, C. (2019). Photocatalytic water splitting of (F, Ti) codoped heptazine/triazine based g-C_3N_4 heterostructure: A hybrid DFT study. *Applied Surface Science*, 463, 809–819.
151. Zhao, Y., Lin, Y., Wang, G., Jiang, Z., Zhang, R., & Zhu, C. (2018). Electronic and optical performances of (Cu, N) codoped TiO_2/ g-C_3N_4 heterostructure photocatalyst: A spin-polarized DFT+ U study. *Solar Energy*, 162, 306–316.
152. Wirth, J., Neumann, R., Antonietti, M., & Saalfrank, P. (2014). Adsorption and photocatalytic splitting of water on graphitic carbon nitride: A combined first principles and semiempirical study. *Physical Chemistry Chemical Physics*, 16(30), 15917–15926.
153. Pan, J., Wang, P., Wang, P., Yu, Q., Wang, J., Song, C., Zheng, Y., & Li, C. (2021). The photocatalytic overall water splitting hydrogen production of g-C_3N_4/CdS hollow core–shell heterojunction via the HER/OER matching of Pt/MnO_x. *Chemical Engineering Journal*, 405, 126622–126641.
154. Xiang, Q., Li, F., Zhang, D., Liao, Y., & Zhou, H. (2019). Plasma-based surface modification of g-C_3N_4 nanosheets for highly efficient photocatalytic hydrogen evolution. *Applied Surface Science*, 495, 143520–143531.
155. Zhou, X., Zhao, C., Chen, J., & Chen, L. (2021). Influence of B, Zn, and B-Zn doping on electronic structure and optical properties of g-C_3N_4 photocatalyst: A first-principles study. *Results in Physics*, 26, 104338–104351.

2 Design and Development of IoT-Based PV Cleaning System

Vinay Gupta, Amit Soni, Himanshu Priyadarshi, and Ronit Banerjee

CONTENTS

2.1 Introduction 29
2.2 Description of Proposed IoT-Based PV Module Cleaning System 33
    2.2.1 Arduino IDE Software...35
    2.2.2 Blynk..35
2.3 Experimental Setup of Suggested IoT-Based PV Module Cleaning System 36
2.4 Results and Analysis..36
2.5 Conclusion 40
References..40

2.1 INTRODUCTION

Photovoltaic (PV) technology has received a lot of attention from the industry as well as the academic research community because it is known to be eco-friendly, noise-free, and has a low cost of adoption [1–3]. However, when confronted with a dusty environment, it faces significant challenges in delivering the expected performance efficiency deliverables [4,5]. Wind, temperature, dust, humidity, solar insolation, and hailstorm all have an impact on PV panel performance as shown in Figure 2.1 [6,7]. The primary cause of PV panel output degradation is dust deposition on the PV panel, which results in massive power reductions over the course of a long-term application [8,9]. Dust is a small firm element with a radius of less than 250 µm. Dust elements include soil, sand, bird droppings, hair, and microfibers [10,11]. Dust build-up is influenced by a range of environmental elements such as moisture, temperature, dust storms, and wind, in addition to the size, shape, chemical characteristics, and weight of the dust. Dust deposition can also be exacerbated by nearby human and vehicle activities [12–14]. Cleaning PV modules is necessary due to the rising demand for solar energy [7].

Because of less traffic and less wind-blowing, the majority of the dust deposition happens at night or before the sun rises. The stagnation of small and big dust particles hanging in the air is ideal under these conditions. The temperature of PV cells is low

DOI: 10.1201/9781003258209-3

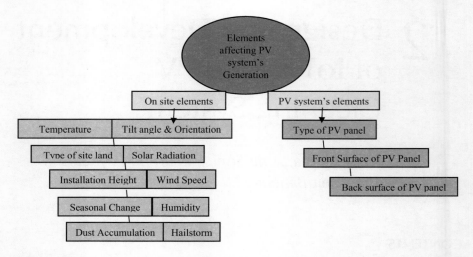

FIGURE 2.1 Elements affecting the PV system's generation.

before the sunrise, and dew forms as water mist in the air condense, interacting with dust particles, increasing cell surface adhesion forces, and making dust deposition layers difficult to clean in a fair manner, resulting in considerable losses in output power generation [15,16].

Figure 2.2 [17] represents the monthly reduction in PV panel efficiency, which in India is greater than 70%. This will have an impact on energy prices, payback periods, and other variables. As a result, regular cleaning of the PV system at a low cost and with limited resources is required.

To mitigate the impact of dust deposition, cleaning techniques and systems have been developed in a number of ways. Table 2.1 shows the comparison of existing cleaning techniques. Depending on the type of cleaning framework, these approaches and methods are costly. These include conventional, manual, and automatic cleaning

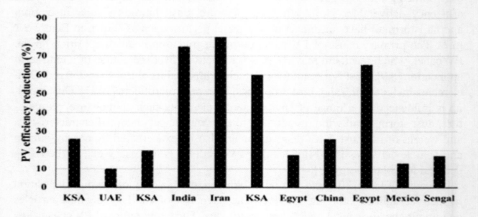

FIGURE 2.2 Monthly reduction in efficiency of PV system due to dust deposition [17].

TABLE 2.1
Comparison of Existing Cleaning Techniques [9]

Cleaning Methods	Natural Cleaning	Manual Cleaning	Mechanical Cleaning	Electrodynamics Screen	Robotic Cleaning	Passive Cleaning
Component	Rain, wind, snow, gravity, tilted panel	Manpower, water, brushes, cloths, detergent, ladder	Wipers, blower, brushes, motors, gears, chains, sensors	High voltage, transparent front screen, sensors	Motor, sensors, brushes, spray, filter	Forming hydrophobic/hydrophilic surface, chemical coating, water
Merits	No cost	Efficient to recover PV performance	Reduces labour cost, automation activation, both cleaning & scrubbing	Very effective and fast cleaning action, no water required, automatic or manual, no mechanical component required, low power consumption	Automatic cleaning and scrubbing, less water required, wireless, rechargeable	No power required, improves the natural cleaning
Demerits	Depends on weather, site dependent, not effective for small dust particles	High cost, water required, scratches may be produced	Maintenance required, power consumption, high capital cost	High capital cost, ineffective for small particles, less durability of transparent screen, performance depends on relative humidity	High cost, slow operation, filter has to be changed periodically	Lifetime is limited, reduces the optical performance
Efficient	Not specific	100%	95%	90%	Not specific	99%
Controlling technique	N/A	N/A	PLC, microcontroller	PLC, microcontroller	Microcontroller, Arduino	N/A

methods. In order to improve the PV module's unwavering quality while reducing the risk of depreciation and financial losses, an effective and efficient mitigation mechanism must be developed.

Manual cleaning is one of the simplest ways to clean PV panels, but this also depends on the water to unsoil the PV panel's surface. This traditional method, in addition to being expensive, necessitates the use of labour. Because water is a rare resource in hot desert conditions and desalination of water produces freshwater, which necessitates a large amount of thermal or electrical energy, this is not a good idea. To clean the PV panel's glass surface manually, the labour uses wipes with proper support structures. Because PV power plants are made up of a number of modules mounted at a height of 10–14 feet or more above the ground, this method is exceedingly time-consuming and difficult.

Hailstorms, in addition to dust accumulation, have an impact on the performance of solar PV panels. A heavy hailstorm can damage the front glass and break the solar cell as shown in Figure 2.3. As shown in Figure 2.4, the PV module has several different types of breaks: diagonal, perpendicular to bus bars, parallel to bus bars, and multi-directional. The generated power of the PV module is significantly reduced by diagonal and multiple directions breaks [18,19]. When cracks appear on the front surface of the glass, they specifically reduce the amount of sun-driven radiation that enters the solar cell. If a solar cell cracks, the current is reduced, and the cell is completely isolated. Hailstorms reduce total electricity generation as well as PV module life; therefore, hailstorm protection for solar PV panels is critical [20].

In this proposed cleaning technique, whenever necessary, a PV cleaning system covers the PV panels and performs the cleaning procedure. As a result, the amount

FIGURE 2.3 Impact of hailstorm on the PV panels [9].

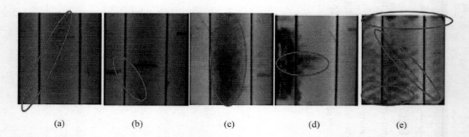

FIGURE 2.4 Types of cracks on the PV panel after hailstorm [9].

IoT-Based PV Cleaning System

of dust collected on the PV surface and the formation of dew is considerably reduced. This chapter describes the design and development of an Internet of Things (IoT)-based PV system cleaning technique that provides an easy-to-use regular cleaning option as well as protection from hailstorms and also experimentally investigates the effectiveness of the proposed PV cleaning system for a solar power plant in a semi-desert environment.

2.2 DESCRIPTION OF PROPOSED IoT-BASED PV MODULE CLEANING SYSTEM

The architecture and control of the proposed IoT-based PV module cleaning system are shown in Figures 2.5 and 2.6. The suggested IoT-based PV module cleaning system includes a DC motor, NodeMCU ESP8266, double relay module, and cleaning brush. At both ends of the PV module, each DC motor is fixed. The flexible wire connects the brush segment to the rotatable pipe. PVC sheet is also rolled on the pipe in such a way that it opens and covers the PV panels when the motor rotates. When it's time to clean and cover the PV panels, users can use the Blynk app on their smartphone to activate the cleaning system. The first DC motor turns after receiving the ON command, moving the brush section and PVC sheet forward. The second DC motor rotates in the opposite direction when the brush segment reaches the PV module's end point, causing the brush segment to go backward. The DC motors are turned off using the reed switch. Table 2.2 shows the electronics and mechanical parts of the suggested IoT-based PV module cleaning system. Table 2.3 shows the specifications of a DC motor used in the experiment. Table 2.4 shows the technical specifications of the electronics and mechanical parts of the suggested IoT-based PV module cleaning system.

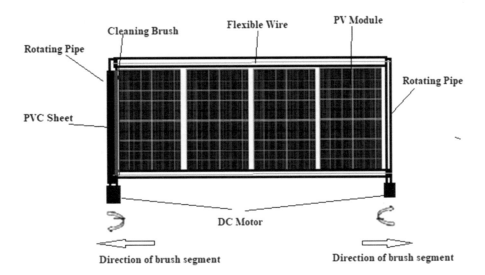

FIGURE 2.5 Proposed IoT-based PV module cleaning system.

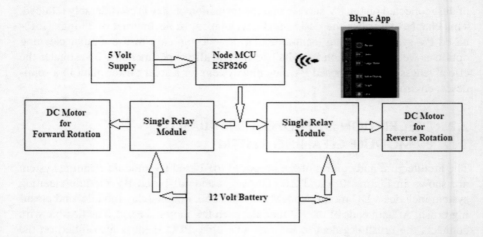

FIGURE 2.6 Block diagram of controller of suggested IoT-based PV module cleaning system.

TABLE 2.2
Electronics and Mechanical Parts of Suggested IoT-Based PV Module Cleaning System

Parts	Hardware/Software	Quantity	Function
Arduino IDE	Software	N/A	For node MCU
Blynk	Software	N/A	To control hardware
Node MCU ESP 8266	Hardware	1	Wi-Fi module
DC motor	Hardware	2	For rotation of brush
Relay module	Hardware	2	To control the switch
Magnetic reed switch	Hardware	2	To stop the DC motor
PVC sheet	Hardware	1	To protect the PV panel
Battery	Hardware	1	For supply
Track rail system	Hardware	N/A	To provide the sliding track for brush
12 Volt supply	Hardware	1	For DC motor
5 Volt supply	Hardware	1	For Arduino
Bearing	Hardware	2	To provide the frictionless support for pipe and motor
Roller	Hardware	2	To provide smoothness during moving of brush pipe on the PV panel
Thread	Hardware	2	To connect sheet to pipe
Cleaning brush	Hardware	1	To clean the PV panels
Mechanical supports	Hardware	2	To connect the motor shaft to study

TABLE 2.3
Specifications of DC Motor Used in the Experiment

Specification	Value	Unit
Type	Gear	-
Input voltage	12	V
Maximum load current	3	A
Rated speed	300	RPM
Stall torque	8	kg-cm

TABLE 2.4
Technical Characteristics of Components used in Proposed System

Name of Component	Parameters	Specification
ESP8266 Wi-Fi module	Supply voltage	3.3 or 5 V
	GPIO Pins	10
	RAM	36 kb
	Clock speed	80/160 MHZ
	MCU	32 bit Ten Sillica L 106
Single relay module	Supply voltage	3.3–5.5 V DC
	Quiescent current	2 mA
	Current range	0–10 A
	Operating temperature	−40°C to 125°C
	Contact voltage range	250 V AC/30 V DC
Reed switch sensor module	Operating voltage	3.3–5 V DC
	Output format	Digital switching output (0 and 1)
	Comparator IC	LM393 IC

2.2.1 Arduino IDE Software

The Arduino IDE is a comprehensive Arduino device emulator software for programming Arduino microcontrollers. The goal is twofold: (i) to link sensing elements, actuating devices, and other types of components and (ii) to use library modules to enable seamless interoperability on both local and global domains.

2.2.2 Blynk

Blynk is an open-source platform that allows users to control and monitor devices in real time using an iPhone operating system or Android mobile. It was designed and built to help the IoT ecosystem by integrating a graphical user interface for back-end control with Raspberry Pi, Arduino, and other devices.

2.3 EXPERIMENTAL SETUP OF SUGGESTED IoT-BASED PV MODULE CLEANING SYSTEM

To test the performance of the proposed IoT-based PV cleaning system, we installed three 20 W PV panels with the proposed cleaning technique on the rooftop of Manipal University Jaipur. Another set of three 20-W PV panels was installed using the traditional method, i.e., fixed. Both sets of PV panels are installed for one month under similar outdoor environment conditions. One set of panels was cleaned and maintained via the suggested IoT-based PV module cleaning system, while the other was not. Using the standard multimeter, the values of short-circuit current and open-circuit voltage of both sets of panels were recorded. The PV solar panels have an inclination of 27° to the south. During the study, commercial PV panels were used with the specifications shown in Table 2.5. Table 2.6 shows the specification of the test location. Figure 2.7 shows the prototype of an IoT-based PV module cleaning system.

2.4 RESULTS AND ANALYSIS

We incorporated three PV panels of 20 W with a proposed cleaning system and another set of three PV panels of 20 W with a conventional method, i.e., fixed, to investigate the performance of the proposed IoT-based cleaning PV system. Both sets

TABLE 2.5
Specifications of Solar Panels

Specification	Value	Unit
Model	ELDORA 20P	-
Type	Polycrystalline	-
Maximum power	20	Watts
Open-circuit voltage	21.44	Volts
Short-circuit current	1.27	Amps
Maximum voltage	17.15	Volts
Maximum current	1.18	Amps

TABLE 2.6
Description of the Test Location

Description	Value
Place	Academic block 1, Manipal University Jaipur
Latitude	26.84
Longitude	75.56
Tilt angle	27°
Facing	South

FIGURE 2.7 Working model of IoT-based PV module cleaning system.

of three polycrystalline solar panels are mounted on a 27° south-facing platform. In this experiment, we installed one set of three PV panels without any cleaning procedure, then applied the proposed cleaning methodology to the second PV set and measured the open-circuit voltage and short-circuit current. The performance of the proposed IoT-based cleaning PV system is determined based on the following points.

i. **Efficient Cleaning**

The suspension of the dust and its deposits in the desert and semi-arid regions, in general, is a natural phenomenon. In addition to accumulating dust, bird droppings, soot, and smoke, other types of organic impurities are seen. The few cells of the PV module are blocked by dust particles such as bird droppings, leaves, and dirt stains.

When there is less wind and less traffic, the majority of dust deposition occurs at night or before the sun rises. Under these circumstances, the stagnation of small and large dust particles suspended in the air is ideal. The temperature of the PV cell is low before sunrise, and dew forms when water mist condenses and interacts with dust particles. The removal of dust deposition layers is difficult, resulting in significant losses in output power generation. The proposed system provides the option to cover the PV panels during the night; so, the adhesiveness of dust can be eliminated.

Real-time test results as shown in Figures 2.8 and 2.9 reveal that the open-circuit voltage and short-circuit current of both PV panels are nearly identical during the early phase. However, the change becomes visible over time a 3.8% difference in open-circuit voltage and a 26% difference in short-circuit current was reported in a month because of dust accumulation on an unclean panel. It demonstrates that the power gain after cleaning is primarily due to a rapid increase in short-current current. The routinely cleaned panel, however, produced consistency in the electrical observables

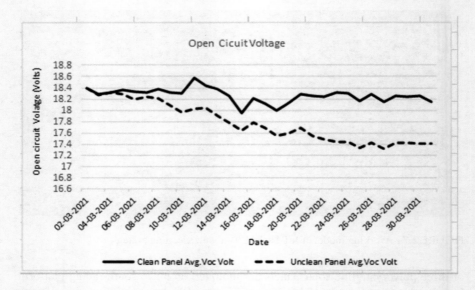

FIGURE 2.8 Comparison of open-circuit voltage between clean and dusty PV panels.

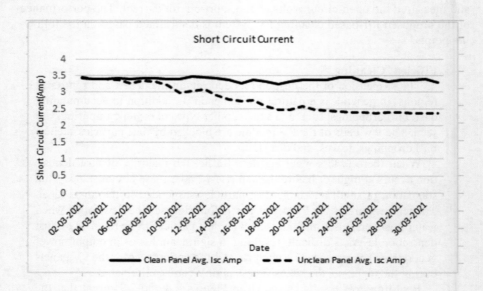

FIGURE 2.9 Comparison of short-circuit current between clean and dusty PV panels.

of interest. The proposed IoT-based PV cleaning system has significantly improved the performance of the PV modules.

ii. **Energy Consumption**

The size and weight of the PV system's characteristics are the primary determinants of the motor's energy consumption, though they are not the only ones. The efficiency of the motors, the battery, and a variety of other factors all have an impact on how much energy they use. While in use, the proposed cleaning method consumes energy. As a result, when evaluating the cleaning method's viability for PV applications, another important factor to consider is the cleaning process' energy consumption. When compared to power gain, this method uses very little energy overall. Table 2.7 shows the average energy consumption during cleaning. The proposed IoT-based cleaning system consumed only 0.478 mWh/day.

iii. **Cost Consideration**

The adaptability of the PV cleaning system depends on its cost. The proposed IoT-based cleaning system is an economical solution for cleaning the PV system. Table 2.8 shows the cost estimation of the proposed cleaning technique with a 60-W PV system. This strategy can be implemented without incurring additional installation costs. The current installation

TABLE 2.7
Average Energy Consumption of Proposed IoT-Based PV Cleaning System

	Average Running Time (s)	Average Current (A)	Energy Consumption (mWh/day)
During forward motion	0.52	0.12	0.208
During backward motion	0.54	0.15	0.270
	Total Energy Consumption 0.478 mWh/day		

TABLE 2.8
Cost Estimation of Proposed IoT-Based PV Cleaning System for 60 W

S. No.	Name of Major Components	No./Length/Area	Rate (Rs.)	Cost (Rs.)
1	Track rail system	50 inch × 24 inch	150 per sq. feet	1,500
2	Roller	2	30 each	60
3	Bearing	1	25 each	25
4	Pipe	6 feet	20 per feet	120
5	Mechanical coupler	1	150	150
6	Cleaning brush	3 feet	30 per feet	180
7	Mechanical supports	1	60 each	60
9	Thread	2 metres	20 per metre	40
10	Protective sheet	12 inch × 24 inch	16 per sq. feet	50
11	DC motor	2	380 each	380
12	Node MCU ESP8266	1	450 each	450
13	5 V supply	1	215 each	215
14	12 V, 2 A supply	1	275 each	275
15	Two relay module	2	90 each	180
Total cost				3,685

procedure has only been slightly altered. Furthermore, no personnel, specialized cleaning equipment, or water is required. It runs on a very low amount of energy. It does, however, gain more energy than it expends. Operating and maintenance costs, such as brush replacement and grease for mechanical parts, are also low.

2.5 CONCLUSION

Several factors influence the PV module's performance, including dust collection, temperature, wind, humidity, and installation location. The dust storm has an adverse effect on the PV system's performance. The dust settles in the solar cell, preventing sunlight from reaching it and lowering the PV system's efficiency. The labour uses wipes with proper support structures to clean the PV panel's glass surface manually. This method is extremely time-consuming and difficult because ground-mounted and roof-mounted PV power plants are made up of a number of modules mounted at a height of 10–14 feet or more above the ground. Hailstorms, in addition to dust accumulation, have an impact on the performance of solar PV panels. A severe hailstorm can destroy the front glass and cause the solar cell to break. PV panels can be permanently damaged by hailstorms. As a result, hailstorm protection for the PV system is essential. The proposed system offers protection against dust storms and hailstorms. The data interpretation clearly demonstrates a small change in the PV panel's open-circuit voltage; however, the short-circuit current is greatly influenced by dust deposition. On the thirty-first day, the dirty PV module's open-circuit voltage and short-circuit current had degraded by 3.8% and 26%, respectively, in comparison to the clean PV panel. The results demonstrate the efficacy of the proposed IoT-based PV module cleaning system. This technique is used to obtain the maximum power out of the PV module while also protecting it from dust storms and hailstorms at a low cost, with low energy consumption, and without the use of water.

REFERENCES

1. Costa, S. C. S., Sonia, A., Diniza, A. C. and Kazmerskia, L. L. (2018). Solar energy dust and soiling R&D progress: Literature review update for 2016. *Renewable and Sustainable Energy Reviews*, 82 (2018), 2504–2536.
2. Zaihidee, F. M., Mekhilef, S. Seyedmahmoudian, M. and Horan, B. (2016). Dust as an unalterable deteriorative factor affecting PV panel's efficiency: Why and how. *Renewable and Sustainable Energy Reviews*, 65 (2016), 1267–1278.
3. Gupta, V. Impact of dust deposition on solar photovoltaic panel in desert region: Review. (2017). *International Journal of Electrical, Electronics and Computer Systems*, 6, 125–140.
4. Gupta, V., Raj, P. and Yadav, A. (2017). Investigate the effect of dust deposition on the performance of solar PV module using LABVIEW based data logger. *2017 IEEE International Conference on Power, Control*, Chennai, India, 2017.
5. Ilse, K. K., Figgis, B. W., Naumann, V., Hagendorf, C. and Bagdahn, J. (2018). Fundamentals of soiling processes on photovoltaic modules. *Renewable and Sustainable Energy Reviews*, 98 (2018), 239–254.
6. Darwish, Z. A., Kazemb, H. A., Sopian, K., Al-Goul, M. A. and Alawadhi, H. (2015). Effect of dust pollutant type on photovoltaic performance. *Renewable and Sustainable Energy Reviews*, 41 (2015), 735–744.

7. Sarver, T., Al-Qaraghuli, A. and Kazmerski, L. L. (2013). A comprehensive review of the impact of dust on the use of solar energy: History, investigations, results, literature, and mitigation approaches. *Renewable and Sustainable Energy Reviews*, 22(2013), 698–733.
8. Sayyah, A., Horenstein, M. N. and Mazumder, M. K. (2014). Energy yield loss caused by dust deposition on photovoltaic panels. *Solar Energy*, 107(2014), 576–604.
9. Gupta, V., Sharma, M., Pachauri, R. K. and Babu, K. D. (2019). Comprehensive review on effect of dust on solar photovoltaic system and mitigation techniques. *Solar Energy*, 191, 596–622.
10. Figgis, B., Ennaouia, A. Ahzia, S. and Rémond, Y. (2017). Review of PV soiling particle mechanics in desert environments. *Renewable and Sustainable Energy Reviews*, 76(2017), 872–881.
11. El-Shobokshy, M. S. and Hussein, F. M. (1993). Effect of dust with different physical properties on the performance of photovoltaic cells. *Solar Energy*, 5(6), 505–511.
12. Kazem, H. A. and Chaichan, M. T. (2016). Experimental analysis of the effect of dust's physical properties on photovoltaic modules in Northern Oman. *Solar Energy*, 139(2016), 68–80.
13. Mani, M. and Pillai, R. (2010). Impact of dust on solar photovoltaic (PV) performance: Research status, challenges and recommendations. *Renewable and Sustainable Energy Reviews*, 14(2010), 3124–3131.
14. Gupta, V., Raj, P. and Yadav, A. (2017). Design and cost minimization of PV analyzer based on arduino UNO. *2017 IEEE International Conference on Power, Control*, Chennai, India, 2017.
15. Hosseini, S.A., Kermani, A.M. and Arabhosseini, A. (2019). Experimental study of the dew formation effect on the performance of photovoltaic modules. *Renewable Energy*, 130, 352e359. Doi: 10.1016/j.renene.2018.06.063.
16. Gupta, V., Sharma, M., Pachauri, R.K. and Babu, K.D. (2019). Impact of hailstorm on the performance of PV module: A review. *Energy Sources, Part A: Recovery, Utilization, and Environmental Effects*, Doi: 10.1080/15567036.2019.164.
17. Kazem, H.A., Chaichan, M.T., Al-Waeli, A.H. and Sopian, K. (2020). A review of dust accumulation and cleaning methods for solar photovoltaic systems. *Journal of Cleaner Production*, 276, ISSN 0959-6526, Doi: 10.1016/j.jclepro.2020.123187.
18. Berardone, I., Corrado, M. and Paggi, M. (2014). A generalized electric model for mono and polycrystalline silicon in the presence of cracks and random defects. *Energy Procedia*, 55, 22–29 Doi: 10.1016/j.egypro.2014.08.005.
19. Dhimish, M., Holmes, V., Mehrdadi, B. and Dales, M. (2017). The impact of cracks on photovoltaic power performance. *Journal of Science: Advanced Materials and Devices*. Doi: 10.1016/j.jsamd.2017.05.005.
20. Kontges, M., Kunze, I., Kajari-Schroder, S., Breitenmoser, X. and Bjørneklett, B. (2010). Quantifying the risk of power loss in PV modules due to micro cracks, In: *25th European Photovoltaic Solar Energy Conference*, Valencia, Spain, September, p. 3745–3752.

3 End Life Cycle Cost-Benefit Analysis of 160 kW Grid-Integrated Solar Power Plant
BSDU Jaipur Campus

Pancham Kumar, Manisha Sheoran,
Anupam Agrawal, Amit Soni, and Jagrati Sahariya

CONTENTS

3.1 Introduction 43
3.2 Solar Photovoltaic Cell 44
    3.2.1 Photovoltaic Energy Conversion 44
    3.2.2 Important Characteristics of Solar Cells 45
    3.2.3 Solar Cells, Module, and Arrays 48
3.3 Methodology 49
3.4 Conclusion 54
References 54

3.1 INTRODUCTION

Solar energy is a leading renewable energy source, and its long-term availability and immense potential have aided in the recovery of the world's developing environmental challenges. Photovoltaic (PV) materials have recently piqued the scientific community's interest in the development of solar cells with a direct bandgap and high absorption co-efficient capabilities. The development of low-cost solar cell materials helps to lower the overall installation cost, making it more affordable to the average person.

    The photovoltaic (PV) effect produces direct electricity from sunlight, where the terms light and electricity are connected to the terms photo and voltaic, respectively. The goal of a solar cell, which is a semiconductor light-sensitive device, is to create direct electricity from sunlight.

The PV effect is a mechanism for producing direct electricity from sunlight, where the terms light and electricity are related to the terms photo and voltaic, respectively. The purpose of a solar cell, which is a semiconductor light-sensitive device, is to generate direct electricity from sunlight. In 1839, Edmund Becquerel was the first French scientist to notice the PV effect [1], but it was only with the development of quantum theory and solid-state physics that it gained its unique character. In 1954, Chapin et al. [2] at Bell Laboratories discovered the first crystalline silicon solar cell with a 6% efficiency. Following that, researchers continued to make every effort to develop highly efficient solar PV materials [3–13]. The United States introduced the first commercially available solar PV system in the 1950s as part of its space program. Although silicon-based PV cells currently account for the majority of the worldwide market, there are other semiconductor materials that, depending on their performance and cost, are offering reasonable competition to Si-based solar cells. The major types of Si and non-Si-based PV materials will be discussed in this chapter.

3.2 SOLAR PHOTOVOLTAIC CELL

When light is incident on a device known as a solar cell, a voltage generating phenomenon (PV effect) occurs. The device utilized in solar cell fabrication is a large-area semiconducting device that collects maximum solar energy from the solar spectrum and converts it into usable electrical energy.

The solar spectrum shows the strength of sunlight as it fluctuates throughout time. The relation between energy (E) and wavelength of radiation is (λ) is:

$$E = \frac{hc}{\lambda} \tag{3.1}$$

where E is the photon energy, h is the Planks constant, and c is the speed of light.

Figure 3.1 depicts the relationship between power (kWm) and wavelength (μm). Figure 3.1 shows that as the wavelength increases, the power decreases after the maximum power is reached.

The total intensity of incoming solar radiation can be calculated from the area under the curve [1,4].

3.2.1 Photovoltaic Energy Conversion

As previously stated, PV energy conversion converts electromagnetic energy associated with ultraviolet, visible, and infrared wavelengths into direct electrical energy (either current or voltage) [5]. The PV energy conversion process is broken down into the following steps:

1. The first step in which absorption of light leads to the transition inside the material from the ground state to the excited state.

End Life Cycle Cost-Benefit Analysis

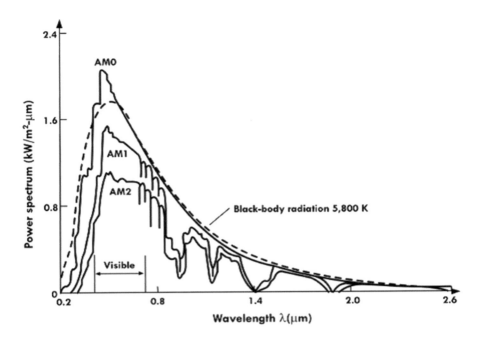

FIGURE 3.1 Solar spectrum associated with various air mass conditions. The spectrum AM0 is related to outside the atmosphere, AM_1 is at the zenith and AM_2 is at an angle of 60°. (Taken from physics of semiconductor devices [22].)

2. In the next step, negative and positive charge carriers are separated in opposite directions via an inbuilt electric field in the depletion region.
3. After this free charge carriers, i.e., negative charge and the charge carriers, are collected on cathode anode contact.

The cross-sectional perspective of a solar PV system is shown in Figure 3.2 to decrease reflectivity losses, and an anti-reflection coating is used.

3.2.2 Important Characteristics of Solar Cells

The electronic properties of a solar PV cell are explored through their equivalent circuit as shown in Figure 3.3.

Diodes are stacked parallel as a current source for an ideal solar cell. Shunt (Rsh) and series resistance (R_s) are introduced in the comparable circuit due to non-ideality of the device.

Total current can be calculated by subtracting photo-produced current from diode current under solar irradiation. Mathematically,

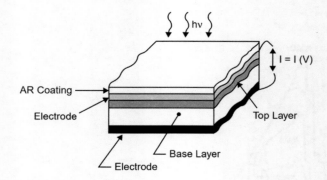

FIGURE 3.2 Cross-section of a typical solar cell.

$$I = I_0(e^{qv/kT} - 1) - I_{ph} \tag{3.2}$$

To understand solar cell characteristics short-circuit current (I_{SC}) and open-circuit voltage (V_{OC}) are two important aspects. These two important situations are shown in Figures 3.4 and 3.5. The charges obtained by solar cell I_{SC} are allowed to move freely through the circuit and no build-up potential is present. Because no diode current is produced in this circumstance, the current generated by the solar cell is at its maximum.

In the case of V_{OC} charge builds up on each side of diode, cause diode current to flow. When the diode current and photo-produced current (I_{Ph}) are equal, the solar cell is said to be in equilibrium. Silicon solar cells typically have a current of 1 mA and a voltage of 0.5 V. The IV characteristics of solar cells are given in Figure 3.6.

The output power of the solar cell is obtained by multiplying current and voltage. To check the quality of solar cells, we use the term fill factor (ff) which is defined as the ratio of maximum power ($P_{max} = I_m * V_m$) to the product of the open-circuit voltage (V_{OC}) and the short-circuit current (I_{sc}).

$$ff = \frac{I_m V_m}{I_{sc} V_{oc}} \tag{3.3}$$

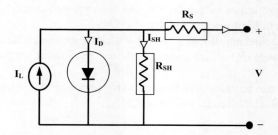

FIGURE 3.3 Equivalent circuit of the solar cell.

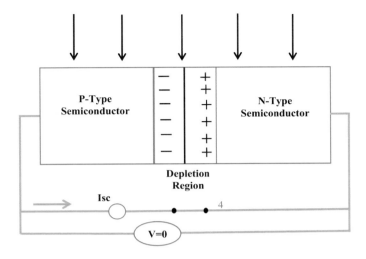

FIGURE 3.4 Short circuits current.

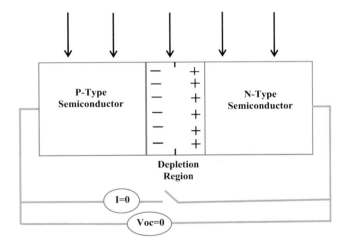

FIGURE 3.5 Open-circuit voltage.

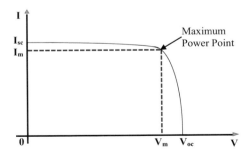

FIGURE 3.6 IV characteristics of the solar cell.

For a perfect rectangular-shaped IV curve, the fill factor value is 1, but generally it ranges between 0.75 and 0.85. Further efficiency of solar cell is calculated as the ratio of output and input power:

$$\eta = \frac{P_{out}}{P_{in}} \tag{3.4}$$

3.2.3 Solar Cells, Module, and Arrays

To make a solar module, solar PV cells are joined in series and parallel. Individual cells are coupled in series and/or parallel to achieve high power levels. These modules are the core building blocks of a PV system. A solar PV array is formed when a number of solar PV modules are joined in series and parallel. A solar PV array can function as a full-fledged power plant. Figure 3.7 (a, b, c) [6,7] shows the whole construction of PV cells, PV modules, and PV array.

Solar energy is a game-changing technology that can help meet current and future global energy demands. The Indian government is taking considerable measures to combat global warming and non-renewable pollutants [14–21]. Solar energy has emerged as a potential option for fulfilling future energy demands [21].

On the roof of Bhartiya Skill Development University Jaipur a 160 KW grid-integrated solar PV plant has been installed. BSDU is India's first skill-based university located in Mahindra World City, Jaipur, Rajasthan. Even if the goal is to provide electricity to every corner of India, the electricity cost still varies in Rajasthan. Using and installing grid-connected solar PV systems can solve this issue. It creates surplus power rather than meeting load requirements. When the same grid (on-grid) generates less power than the load demand, it takes the surplus electricity from the grid to meet the load. To promote easy recycling and eliminate dangerous components, manufacturers should be held accountable for end-of-life standards. This will aid in the development of an autonomous and sustainable society in the face of the fossil-fuel crisis.

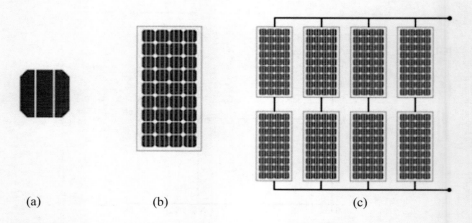

(a)   (b)   (c)

FIGURE 3.7 (a) Solar PV cell. (b) Solar PV module. (c) Solar PV array.

3.3 METHODOLOGY

In April 2017, 160KW on-grid rooftop solar PV plant was installed at BSDU Jaipur campus. The site consists of 500 numbers of Anchor Panasonic crystalline silicon panels of 320 Wp rating. The solar panel, inverter, battery, mount structure, and storage batteries are all part of this system. The panels at the rooftop grid-connected plant are fixed at a 0° azimuth angle and a 26° inclination leaning south. Panels are installed in a shade-free location during installation. Anchor Panasonic crystalline silicon panels rated at 320 Wp are used. The mechanical characteristics of the planned solar plant's solar panels are listed in Table 3.1. These panels are resistant to elements such as salt, moisture, dust, rust, and shadow.

On-grid solar PV system designing of BSDU 160 kW on-grid solar power plant has been performed. For string sizing inverter datasheet and solar PV module datasheet has been utilized. For proper string sizing approximate maximum (i.e., 45°C) and minimum (10°C) BSDU Jaipur site location temperature has been also taken into account.

There are a total of 500 320-Watt Anchor Panasonic make solar PV modules and three 50 kW, 3ϕ, 400V ABB make inverters are used to construct 160 kWp grid-connected solar PV system. The technical datasheet of solar PV modules and inverters are shown in Tables 3.1 and 3.2, respectively.

Designing Steps of 160 kW BSDU Solar PV Plant:
Estimate Required Number of PV Modules

PV modules with various power ratings (e.g., 3 Wp, 6 Wp, ..., 100 Wp, 200 Wp, 300 Wp) are available on the market.

TABLE 3.1
Technical Datasheet of ABB Make 50 kW on Grid Inverter

Absolute maximum DC input voltage (V_{max})	1,000 V
Start-up DC input voltage (V_{start})	300–500 V (default 360)
Rated DC input voltage V_{dcr}	715 rated DC input V_{dc}
Rated DC input power P_{dcr}	51,250 W
Number of independent MPPT	1
MPPT input DC voltage range	520–800 V_{dc}
Maximum DC input current (I_{dcmax})	100 A
Maximum input short-circuit current	144 A
Number of DC inputs strings/pairs	12 or 16 string combiner version available
AC grid-connection type	3-Phase, 440 V
Rated AC power	50,000 W
Maximum AC output power	50,000 W
AC voltage range	422–528 V
Maximum AC output current (I_{acmax})	61 A
Rated output frequency (f_r)	50 Hz

TABLE 3.2
Technical Datasheet of 320 Watt Anchor Panasonic Make Solar Panel

STC	AE7P320VB4B	AE7P315VB4B	AE7P310VB4B
Maximum power at STC (P_{max})	320 W	315 W	310 W
Optimum operating voltage (V_{mp})	37.1 V	36.8 V	36.5 V
Optimum operating current (V_{mp})	8.63 A	8.56 A	8.50 A
Open-circuit voltage (V_{OC})	45.6 V	45.1 V	44.9 V
Short-circuit current (I_{sc})	9.14 A	9.02 A	8.96 A
Module efficiency	16.5%	16.2%	16.0%
Operating module temperature		−40°C to +85°C	
Maximum system voltage		1,000 V DC	
Module weight		25.8 kg	
Maximum series fuse rating		20 A	
Temperature coefficient of V_{OC}		0.33%/ °C	

We utilize Panasonic Anchor make 320 Wp PV modules that are readily accessible on the market. For a 160 kW solar PV system, the total number of PV modules required would be:

$$= \frac{160 \text{ kW}}{320 \text{ Wp}} = 160 \text{ kW}/320 \text{ Wp} = 500 \text{ modules}$$

Hence, 500 number of solar PV modules will be required for a 160 kW plant.

Effect of Temperature on Voltage
Temperature coefficient of voltage = −0.33%/°C
Voltage rise in winter = $V_{OC} \times (-0.0033)$ (STC temperature − Minimum temperature)

$$= 45.6 \times 0.0033 \ (25 - 15) = 1.50 \text{ V}$$

Total voltage in winter/Maximum voltage = V_{OC} + 1.50 V = 45.6 + 1.50 = 47.1 V
Voltage fall in summer = $V_{OC} \times (0.0033)$ (STC temperature − Minimum temperature)

$$= 45.6 \times 0.0033 \ (25 - 45) = -3.00 \text{ V}$$

Total voltage in summer/Minimum voltage = V_{OC} −3.00 V = 45.6−3.00 = 42.6 V

Maximum and Minimum Number of Modules in a String

$$\text{Minimum modules in a string} = \frac{\text{Inverter MPPT lower start voltage}}{\text{PV Module } V_{mp}}$$

$$\frac{360}{47.1} = 7.64 = 8 \text{ modules}$$

$$\text{Maximum modules in a string} = \frac{\text{Maximum MPPT DC input voltage of inverter}}{\text{PV Module } V_{mp}}$$

$$= \frac{800}{47.1} = 16.98 = 17 \text{ modules}$$

Number of strings in parallel = Total number of PV modules in the system/Number of modules in a string $= \frac{500}{17} = 29.41 = 30$ parallel strings (by considering maximum modules in the string) $= 29.41 = 30$ parallel strings (by considering maximum modules in the string)

Maximum and Minimum Number of Modules in a String

Since a maximum of 17 modules are used to form a series string, their output voltage is

$$= 17 \times V_{OC}$$

$= 17 \times 47.1 = 800 < 800$ V (i.e., equal to maximum MPPT DC input voltage of inverter accepted)

Hence, 30 parallel strings of 17 modules are possible. Here we are using three ABB make, 50 kW, 3-phase, 440 V rating inverters.

$$\text{Total number of parallel string per inverter} = \frac{\text{Maximum number of parallel strings}}{\text{Number of inverters}}$$

$$= \frac{30}{3} = 10 \text{ parallel strings per inverter (approximately)}$$

So we can connect a maximum of 10 number of parallel strings per inverter
Total output current of the parallel string

$$= \text{No. of parallel string} \times \text{Maximum current of a module}$$

$= 10 \times 9.14 = 91.4 < 100$ A (Maximum DC input current of inverter accepted)

The interconnection diagram of BSDU 160 kW solar power plant is shown in Figure 3.8.

Results Discussion

Table 3.3 shows that the total unit generation of the 160 kW solar power plant on the BSDU Jaipur campus was 245,876 kWh from April 2018 to March 2019. The cost per unit electricity consumption is Rs. 8.35.

Total savings from plant (April 2018 to March 2019) $= 245,876 \times 8.35 =$ Rs. 2,053,064

The materials and installation cost of the BSDU 160 kW solar PV plant is around Rs. 8,467,500, as shown in Table 3.4.

The plant has a 25-year life cycle. We can compute the plant's energy payback period to see how long it will take for the plant to provide free electricity.

Energy payback Period = Total cost of solar power plant/yearly unit cost generation by solar plant = 8,467,500/2,053,064 = 4.12 years

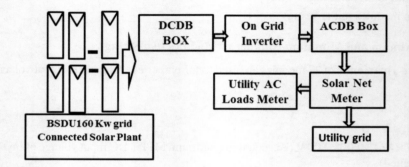

FIGURE 3.8 BSDU 160 kW solar power plant.

TABLE 3.3
Main Solar Meter Reading

Month	Date of Reading	Total kWh
April	1-04-2018	24,240
May	1-05-2018	23,500
June	1-06-2018	24,144
July	1-07-2018	21,020
August	1-08-2018	16,856
September	1-09-2018	15,652
October	1-10-2018	17,364
November	1-11-2018	22,108
December	1-12-2019	19,344
January	1-01-2019	20,188
February	1-02-2019	20,140
March	1-03-2019	21,320
Total kWh		245,876
Monthly average kWh		20,489

TABLE 3.4
BSDU 160 kW Solar PV Plant Materials & Installation Cost

S. No.		Description	Amount (INR)	Rate Per/Watt
Materials		Solar PV modules	8,136,000	54.24
		Inverters		
		MS galvanized structure		
		Balance of system components: AC/DC wires, light arrestor, PVC, conduits, earthing		
Installation and commissioning charges		Installation and commissioning charges	3,31,500	2.21
Total cost			8,4,67,500	

Energy payback Period = Total cost of solar power plant/yearly unit cost generation by solar plant = 8,467,500/2,053,064 = 4.12 years

Since the life cycle of a plant is around 25, years, and after 4.12 years the plant will generate free electricity for approximately the next 21 years.

After 4.12 years, for the next 21 years electricity generated by solar plants is almost free.

Total free electricity produced by solar plant (A) = 21 × 2,053,064 = Rs. 43,114,344

End Life Solar PV Module Cost Analysis

When the solar panels on the 160 kW solar power plant approach the end of their useful life, a significant amount of garbage will be generated. This can be handled by either recycling the panels via a thermal, mechanical, or chemical technique, or by removing the panels [23]. It is also possible to calculate the benefit derived from the solar plant at the end of its useful life based on the market price of recycled and dismantled products [24]. D'Adamo and colleagues [25] estimated the mass fraction of tons of crystalline silicon modules. Following the end of a solar plant's usable life, the amount of money generated by the solar panel is computed mathematically in this section.

The BSDU 160 kW solar power plant is made up of 500 number of solar modules, each weighing 25.8 kg.

The total weight of 500 modules is 12,900 kg (500×25.8) (i.e., 12.9 ton). The recovery yield from the 160 kW solar PV plant is 100% for Al, 97% for glass, 85% for Si, 78% for Cu, and for plastics include the ethylene vinyl acetate sheet, which is not included in the calculation due to its fractional recovery compared to others [26]. Table 3.5 shows the composition of materials after recycling.

Market price of each recovered element is investigated [14], and it is found that the total recovered amount is around Rs. 666,240 as mentioned in Table 3.6.

As a consequence, a total of Rs. 666,240 in gross profit was saved. However, only the cost of solar module recycling is considered; if the cost benefit of materials recovered from dismantling the entire balance of system components (i.e., structure, connectors, cables, ACDB & DCDB boxes, inverters, and so on) is also considered, the cost benefit would be much greater.

TABLE 3.5
Material Composition after Recycling

S. No.	Recycled & Recovered Materials	% Material Composition	Material Mass Recovered/12.9 Ton (in Ton)	Material Mass Recovered/12.9 Ton (in kg)
1	Al	17.5	2.257	2,257
2	Glass	65.8	8.488	8,488
3	Silicon	2.9	0.3741	374.1
4	Copper	1	0.129	129
5	Plastic	12.8	1.652	1,652

TABLE 3.6
Composite Material Costing after Recycling

S. No.	Recycled & Recovered Materials	Material Mass Recovered/109.37 Ton (in kg)	Market Price (Rs.)/kg	Price of Recycled Material
1	Al	2,257	145.12	327,535
2	Glass	8,488	9.07	76,986
3	Silicon	374.1	126.98	47,503
4	Copper	129	444.43	57,276
5	Plastic	1,652	95	156,940
	Total Cost(B)			6,66,240

So, total plant saving during and after their life cycle = A + B = 43,114,344 + 666,240 = Rs. 43,780,584.

3.4 CONCLUSION

The current research looks at the design and lifecycle assessment of a 160 kW rooftop solar PV system. This includes significant power bill reductions, as system usage increases any institution's financial burden. The overall savings from a 160 kW BSDU solar PV plant during its life cycle and beyond its end-of-life cycle is Rs. 43,780,584, which is a significant amount of money saved in terms of electricity costs. After the energy recovery period is met, the complete reward is the free electricity generation.

REFERENCES

1. A.E. Becqurel, "On electric effects under the influence of solar radiation". *C.R. Acad. Sci.*, 9 1839, 561–567.
2. D.M. Chapin, C.S Fuller, and G.L. Pearson, "A new silicon p-n junction photo cell for converting solar radiation in to electrical power". *J. Apply. Phys.*, 25 1954, 676–677.

3. A.L. Faharenbruch, V. Vasilchenko, F. Buch, K. Mitchell, and R.H. Bube, "II-VI photovoltaic hetrojunctions for solar energy conversion". *Appl. Phys. Lett.*, 25 1974, 605–608.
4. A. Tubtimate, K.L. Wu, H.Y. Tung, M.W. Lee, and G.J. Wang, "Ag2S quantum dot sensitized solar cells". *Electrochem. Commun.*, 12 2010, 1158–1160.
5. J.D. Mcgee, "Photoelectric cells-A review of progress". *IEE Proc. part B: Radio Electron. Eng.*, 104 1957, 467–484.
6. M.A. Green, "Silicon solar cells: Evolution, high efficiency design and efficiency enhancements". *Semicond. Sci. Technol.*, 8 1993, 1–12.
7. J. Appelbaum, M. Shechter, J. Bany, and G. Yekutieli, "Array representation of non identical electrical cells". *IEEE Trans. Elect. Dev.*, 29 1982, 1145–1151.
8. Irena working paper, *Renewable Energy Technologies: Cost Analysis Series*. IRENA, Germany, 2012.
9. M.A. Green, *Clean Energy from Photovoltaics*. World Scientific Publishing Co., Hackensack, NJ, 2001.
10. J. Szlufcik, S. Sivoththaman, R. P. Mertens, and R.V. Overstraeten, "Low-cost industrial technologies of crystalline silicon solar cells". *Proc. IEEE*, 85 1997, 711–730.
11. M.A. Green, Y. Hishikawa, E.D. Dunlop, D.H. Levi, J.H. Ebinger, and A.W.Y. Ho-Baillie, "Solar cell efficiency tables (version 51)". *Prog. Photovolt Res. Appl.*, 26 2018, 3–12.
12. Dena German Energy Agency, "Information about German renewable energy, industries, companies and product (Federal Ministry of Economics and Technology)". 2013–2014, 41.
13. J. Szlufcik, S. Sivoththaman, J.F. Nijs, R.P. Martens, and R.V. Overstraeten, "Low cost industrial technologies of crystallines silicon solar cells". *Proc. IEEE*, 85 1997, 711–730.
14. A.K. Shukla, K. Sudhakar, and P. Baredar, "Simulation and performance analysis of 110 kWp grid-connected photovoltaic system for residential building in India: A comparative analysis of various PV technology". *Energy Rep.*, 2 2016, 82–88.
15. R. Khatri, "Design and assessment of solar PV plant for girls hostel (GARGI) of MNIT University, Jaipur city: A case study". *Energy Rep.*, 2 2016, 89–98.
16. V. Suresh, and S. Sreejith, "Economic dispatch and cost analysis on a Power system network interconnected with Solar Farm". *Int. J. Renew. Energy Res. (IJRER)*, 4 2015, 1098–1105.
17. F. Farzan, S. Lahiri, M. Kleinberg, K. Gharieh, F. Farzan, and M. Jafari. "Microgrid for fun and profit". *IEEE Power Energy Mag.*, 2013, 52–58.
18. S. Brini, H.H. Abdallah, and A. Ouali, "Economic dispatch for power system including wind and solar thermal energy". *Leonardo J. Sci.*, 14 2009, 204–220.
19. Y.-K. Wu, G.-T. Ye, and M. Shaaban, "Analysis of impact of integration of large PV generation capacity and optimization of PV capacity: Case studies in Taiwan". *IEEE Trans. Indus. Appl.*, 6 2016, 4535–4548.
20. F. Ferroni, and R.J Hopkirk, "Energy return on Energy invested (ERoEI) for photovoltaic solar system in regions of moderate insolation". *Elsevier Energy Rep.*, 94 2016, 336–344.
21. K. Singh, S. Mishra, and M.N. Kumar, "A review on power management and power quality for islanded PV microgrid in smart village". *Ind. J. Sci. Tech.*, 10 2017.
22. S.M. Sze, Y. Li, and K.K. Ng, *Physics of Semiconductor Devices*. National Chiao Tung University, Taiwan, 2006.
23. A. Domínguez, and R. Geyer, "Photovoltaic waste assessment in Mexico". *Res., Conserv. Recycl.*, 127 2017, 29–41.
24. P. Dias, and H. Veit, "Recycling crystalline silicon photovoltaic modules". *Emerg. Photovolt. Mater.: Silicon Beyond*. John Wiley Sons, 2018, 61–102.

25. I. D'Adamo, M. Miliacca, and P. Rosa, "Economic feasibility for recycling of waste crystalline silicon photovoltaic modules". *Int. J. Photoenergy*, 2017, 1–6.
26. Aluminium Price, "MCX Aluminium - Aluminium Rate in India Today Live on The Economic Times". *The Economic Times*. https://economictimes.indiatimes.com/commoditysummary/symbol-ALUMINIUM.cms (accessed Oct. 09, 2020).

4 A Review of the Theoretical Results Associated with the Intermediate Bandgap Solar Cell Materials
A Density Functional Study

*Aditi Gaur, Karina Khan, Amit Soni,
Jagrati Sahariya, and Alpa Dashora*

CONTENTS

4.1	Introduction	58
	4.1.1 First-Generation Photovoltaic Cells	59
	4.1.2 Second-Generation Photovoltaic Cells	60
	4.1.3 Third-Generation Photovoltaic Cells	60
4.2	Thin-Film Solar Cells	61
4.3	Experimentally Studied Thin-Film Solar Cells	64
	4.3.1 Binary Thin-Film Solar Cell	65
	4.3.2 Ternary Thin-Film Solar Cell	65
	4.3.3 Quaternary Thin-Film Solar Cell	66
4.4	Theoretical Thin-Film Solar Cells	66
	4.4.1 Binary Thin-Film Solar Cell	67
	4.4.2 Ternary Thin-Film Solar Cell	68
	4.4.3 Quaternary Thin-Film Solar Cell	69
4.5	Bulk Intermediate Band Solar Cell	70
4.6	Thin-Film Intermediate Band Solar Cell	71
4.7	Conclusion	73
References		73

4.1 INTRODUCTION

With the rapid growth of population and technologies, the requirement for energy increases at a tremendous level which is majorly fulfilled through conventional sources that include fossil fuels such as coal, natural gas and petroleum. The overexploitation of fossil fuels become the prime concern for environmentalists because their nature is exhaustible and their combustion causes air pollution too [1]. The limiting nature of these conventional sources makes it an urgent need to adopt renewable and pollution-free sources which can compete with the efficiency produced by the present sources. Renewable energy includes solar, wind, hydro, waves and geothermal as they are derived from the renewable resources that are replenishable themselves [1,2]. Among all the available renewable sources, sunlight provides a huge amount of solar energy on which other sources are also dependent, and it is the most authentic source that is convenient to use and maintain. Solar energy is used directly to maintain life on the earth but to make this energy into a useful form like fossil fuels, it is important to harness solar energy that can be achieved using photovoltaic modules. The energy conversion through photovoltaic modules refers to converting solar energy into the electrical form that can be utilized directly or can be stored. The photovoltaic cells are made up of a semiconductor diode that can soak sunlight in the form of photons and transfer this energy to electrons to create electron-hole carriers which generate electric current. Figure 4.1 shows the basic components of photovoltaic cells in which it is presented that the sun rays in the form of photons fall on the front part of the cell where the photo-absorbing material is installed and a grid is used to make electrical contact between the diode where the photons are absorbed and converts into the electrical form [4]. Moreover, for obtaining more efficiency, photovoltaic cells are being

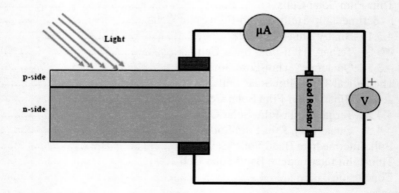

FIGURE 4.1 Basic diagram of photovoltaic cell [3].

TABLE 4.1
Classification of Photovoltaic (PV) Cell among Different Generations

I Generation PV Cells	II Generation PV Cells	III Generation PV Cells
Monocrystalline Si PV cell	Thin-film amorphous Si cell	Dye-sensitized PV cell
Polycrystalline Si PV cell	CdTe thin-film PV cell	Perovskite PV cell
Amorphous silicon PV cell	CIS/CIGS thin-film PV cell	Organic/polymer PV cell

continuously developed by the research community. Along with this, the materials' abundance, non-toxicity, cost of material and installation, durability, stability at high temperature, etc. become the important parameters for their gradual development [5,6]. Based on the inventions, these photovoltaic cells are bifurcated into three generations each: first-generation photovoltaic cells, second-generation photovoltaic cells and third-generation photovoltaic cells, and their further classification are collated in Table 4.1.

4.1.1 First-Generation Photovoltaic Cells

The oldest photovoltaic technology belongs to the first generation which came into existence in the early 18th century. The conventional cells fall and account as the most frequently used domestic cells among all the generations of photovoltaic cells as they are made up of thin wafers of silicon which offers good efficiency and tenability. Silicon is the second most ample source in the semiconductor element category on earth and its bandgap is 1.1 eV, so its utilization as the photovoltaic cell is considered as most appropriate. Based on the manufacturing of silicon wafers, the first-generation crystalline solar cells are categorized into four types: monocrystalline, polycrystalline, amorphous and hybrid silicon photovoltaic cells [5]. Monocrystalline photovoltaic cells are made by slicing the large single crystal of silicon and their installation is done on the compact area. These cells have efficiency of up to 24% but this skittles away with an increase in temperature. Polycrystalline photovoltaic cells are the impure form of monocrystalline photovoltaic cells and comparatively of lower cost. These cells are manufactured by creating multiple wafers through the molten silicon and their recorded efficiency is up to 21%. Amorphous photovoltaic cells are composed by forming a thin layer of silicon on the substrate but these cells recorded very little efficiency and are not eligible to work in slightly higher temperatures. The only advantage of amorphous photovoltaic cells is that they exhibit flexibility and can easily be used in small devices. Hybrid photovoltaic cells are made by hybridization of amorphous and monocrystalline cells; they recorded better efficiency which is maintained at a certain high temperature and with this they can easily work on indirect incident light [1–5].

4.1.2 Second-Generation Photovoltaic Cells

Thin-film photovoltaic cells emerged in this generation of photovoltaic cells in which a thin layer of a few micrometers is formed on the substrate. These solar cells beat crystalline-based photovoltaic cells in many aspects; these provide much better efficiency with lower cost and being lightweight, along with flexibility they are easy to install. Commercially, there are three types of primary thin-film photovoltaic cells: amorphous silicon (α-Si), cadmium telluride (CdTe), copper indium selenide (CIS) or copper indium gallium diselenide (CIGS) thin-film photovoltaic cells. Thin-film amorphous silicon cells are designed with p-n junction and thin film used in this type of cells have high bandgap to captivate the highest amount energy of photons and its lower layer is comprised of a material having lower bandgap; this combination allows the second-generation amorphous silicon to give better efficiency than first-generation PV cells. Cadmium telluride thin-film photovoltaic cells are considered as the stable compound which is manufactured by sandwiching the cadmium telluride p-type layer into n-type cadmium sulfide layer that results in 1.4 eV bandgap which is very near to the ideal photovoltaic cell's bandgap. In addition, these thin-film photovoltaic cells have comparatively better efficiency than amorphous silicon photovoltaic cells.

Copper indium selenide thin film and copper gallium selenide thin film with doping of sodium form copper indium selenide and copper indium gallium diselenide thin-film photovoltaic cells [5,7]. These are direct bandgap photovoltaic cells, which are comparatively low-cost photovoltaic cells, and environment friendly as cadmium-based cells are not eco-friendly. These cells exhibit good efficiency (21.6%) and can work if the temperature increases [5].

4.1.3 Third-Generation Photovoltaic Cells

These are improved and the latest technology among all photovoltaic cells that includes nanotubes, silicon wires and dye-sensitized, quantum dots photovoltaic cells. This generation has emerged to lower the cost and toxicity that are faced in both previously developed photovoltaic cells. Generally, the third generation includes dye-sensitized, perovskite and organic photovoltaic cells [5,8–10]. Dye-sensitized photovoltaic cells are composed of dye sensitizer, electrolyte, electrode and nanocrystalline semiconductors. These photovoltaic cells are considered the best ones due to their high efficiency, low fabrication cost and eco-friendly feature. In addition, their output depends on the thickness of the electrode and the nature of the dye used. However, the performance of dye-sensitized photovoltaic cells are affected by the organic solvent used in it, which easily evaporates [7,8]. Perovskite solar cells belong to the ABX_3 family. As per the research in the last few years, these cells have gained attention because of the feasible solution they provide in terms of production cost, tremendous absorption of light and highly efficient results [11].

Organic photovoltaic cells are also known as polymer photovoltaic cells because their photon interacting layer is made up of polymers that offer maximum flexibility. These cells offer comparatively low efficiencies but are manufactured at a very low cost [5].

4.2 THIN-FILM SOLAR CELLS

The solar cell industry's major dependence for its future market is 'thin-film solar cell'. This is due to their less material requirement, more economical, least toxic components, less waste generation and easy manufacturing nature [12].

So basically, it is a type of solar cell that belongs to the second generation composed by coating many such layers of photovoltaic material or thin-film (TF) on physical substrate layers like plastic, glass or metal. The thin-film solar cells contain three categories within themselves such as α-Si, CdTe and CIGS thin films. These categories along with their formation processes are explained below:

a. **Amorphous silicon (α-Si) thin films**: This is the category in which the composition of a thin film is through an amorphous Si compound that is a non-crystalline silicon used to produce mono or polycrystalline-based solar cells. They have been in the solar cell market for more than 20 years. They are known to be the best established and oldest thin-film silicon technology.
   Formation process: The hydrogenated amorphous silicon has been formed by the processing of glass as substituent material at a temperature lower than 600°C and is deposited by the plasma-enhanced chemical vapor deposition. This technique has been utilized as the baseline thin-film photovoltaic technology processing at 200°C [13]. As per the reports from 1975, hydrogenated amorphous silicon has been noted to have behaved as a semiconductor. Its conductivity is varied for many magnitude orders by inserting impurities/dopant atoms. The schematic diagram can be seen in Figure 4.2.
   Limitation/alternatives: This methodology offers notable benefits such as economic PV electricity and high optical absorption coefficient (thickness of absorber film 300 nm or lesser). The drawback of this category is the stable average efficiency which is quoted to be equal to 6% or lesser

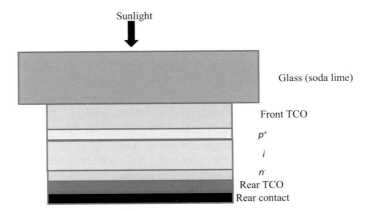

FIGURE 4.2 Schematic of a p-i-n hydrogenated amorphous silicon solar cell present on a glass superstrate [13].

than large-area single-junction PV modules [14]. It had limitations in the material caused due to its meta-stability and disorder but was resolved by the introduction of a large-area plasma-enhanced chemical vapor deposition technique. The better option provided in replacement to this compound is monocrystalline Si thin film which provided a lower bandgap to be utilized as active bottom cell material. But this type had a problem in establishing over a large area and reducing light-induced degradation. Finally, poly-silicon thin-film solar cells provide an alternative to photovoltaic technology.

b. **Cadmium telluride (CdTe) thin films**: CdTe lies among the very first materials to be studied for applications of solar cells. Unique properties for the PV cells and modules are easily offered by CdTe with a direct bandgap of 1.5 eV (as per Shockley–Queisser, it lies in the nest range for solar energy conversion) [15]. It offers high-efficiency device pre-requisites such as sufficient lifetime for charge carrier and mobility.

   Formation process: Deposition of CdTe thin film is done by physical & chemical techniques such as sputtering, close-spaced vapor transport (CSVT), electro-deposition and also spray pyrolysis. A known fact about CdTe films is that thermal treatment of $CdCl_2$ is required to enhance morphology and reduce centers of recombination [16], and another treatment is done with the use of fluorine. The thin-film deposition of CdTe can be achieved by close space sublimation and vapor transport deposition methodology. The technique of evaporating of CdTe compound and the condensation of this compound on a medium cool substrate are included in both the above-mentioned methods. Close space sublimation involves both source and substrate lying in the same section and substrate lies over the source (known as bottom-up evaporation). The vapor transport deposition process makes use of carrier gas to move vapor from the source position to the substrate layer position. It enables top-down configuration having benefits over substrate's mechanical essentials and movement of the substrate through the deposition machine. Figure 4.3 depicts the CdTe formation process.

   Limitation/alternatives: In terms of large-scale production, the scarcity of Te supply turns out to be a drawback for CdTe-based modules and can also raise environmental concerns. It offered a limited efficiency percentage of 16% for the cells that were laboratory based. The solution was provided to eradicate the reduced losses caused by host absorption in the CdS compound and the loss of open-circuit voltage (0.6–0.8 V). Due to this solution's high efficiency, improved open-circuit voltage along with the fill factor was achieved. Solutions like embedding composition gradient provided improved control over doping and reduced contact and interface recombination [17]. Due to this, the efficiency achieved is more than 25% with polycrystalline thin-film solar cells.

c. **Copper indium gallium selenide (CIGS) thin films**: The basic description of CIGS PV technologies begins with the absorber layer of the semiconductor compound where the solar light is absorbed [18]. CIGS

Intermediate Bandgap Solar Cell Materials

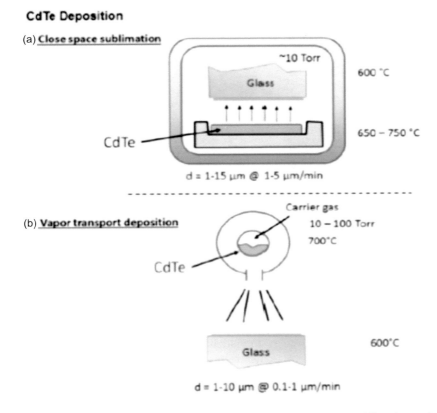

FIGURE 4.3 Formation process for CdTe deposition: (a) close space sublimation and (b) vapor transport deposition [18].

is found to exist in chalcopyrite structure form. Tetrahedral bonding is present in the chalcopyrite crystal structure but there is a variation in the bond lengths with alternative elements resulting in the shift of lattice parameters, thus leading to a change in the bandgap of the semiconductor. The bandgap of the CIGS thin-film layer is defined by a composition that doesn't require to be uniform. Also, the feature of composition gradient and uniformity in the lateral area of solar cells presents a suitable quotient for higher efficiency.

Formation process: In the formation process of such thin films, two methodologies have been discussed, namely (i) sequential and (ii) co-evaporation process. In sequential process, deposition of the needed amount of Cu, In and Ga precursor layers onto the substrate is done. In this step for the process of coating, inks, evaporation, electro-depositions could be used. Although for production, the method which is widely used and is renowned is sputtering. In the case of the co-evaporation process, all the contributing elements (Cu, Ga, In and Se)

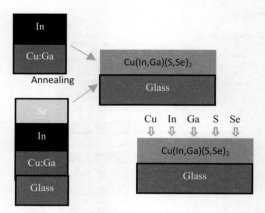

FIGURE 4.4 Deposition method for CIGS: (a) sequential process and (b) co-evaporation process [18].

are evaporated thermally in the condition of the system with high vacuum and also these elements are then condensed onto the heated substrate. Figure 4.4 depicts the deposition method for CIGS.

Limitation/alternatives: The main technical challenge in the CIGS technology is the complex formation of the CIS absorber layer (viz. a five-element system), which raises the issue of realizing uniform film properties over large-area substrates with the use of high-throughput equipment. Due to this, the output produced and the price of the modules is affected (€/Wp). The use of cadmium (Cd) raises the issue of toxicity and indium is considered to be a scarce element in the environment now. It is believed that the known source of indium reserves would only be available for assembly/ manufacturing of a few CIS PV modules (GWp). Many material defects are observed in Cu(In;Ga)Se$_2$ thin-film layers which are present within crystal bulk, at the surface and grain boundaries. The origination of defects occurs from faults in crystal structure: stacking faults, vacancies, anti-site defects and the presence of extra alkali elements. P-type doping is caused due to natural defects such as Cu and Se vacancies which make it difficult to dope CIGS n-type material.

4.3 EXPERIMENTALLY STUDIED THIN-FILM SOLAR CELLS

There have been many research works that define the different methodologies utilized to study the optical properties and other important features of thin-film solar cells. One such kind of work which is a compilation of several other research works based on experimental and simulation studies is explained by Ruiz [19]. They have defined the optical depiction and outlined various kinds of solar cells and thin-film solar cells. Specific tools for optical measurement such as photoluminescence, ellipsometry and photo reflectance are mentioned in it. This compilation presents the

Intermediate Bandgap Solar Cell Materials 65

materials and measurements which are made by analyzing suitable physical principles. Concluding this, they present a numerical model that can be used to design the structure of a thin-film solar cell [19]. Experimental work on the three categories (based on the number of elements in a compound) has been explained below.

4.3.1 Binary Thin-Film Solar Cell

A solar cell is experimentally created using a thin-film and multilayer system of silicon dioxide (SiO_2) and silicon nitride (Si_3N_4) based on the grating. It is developed using the chemical vapor deposition method. The multilayer thin-film structure, grating and both combined present 8%, 21% and 44% enhanced cell efficiency, respectively [20]. The optical performance for the solar cell is decided by the thickness of the active layer and also the grating height. Regarding the planar and SC4 solar cells, enhanced cell efficiency has been offered for the later designed compound to about 79% and 21%. A novel binary thin-film semiconductor material, BiI_3 has been discovered as a potential absorber material in the TFSCs. Two-step methodology has been followed through the iodization of BiSI precursor TF at a low temperature value [21,22]. Binary metal sulfides such as Cu_2S, Sb_2S_3, SnS, Ag_2S, PbS [23], Bi_2S_3 and FeS_2 have proved to be cost efficient in providing absorber material solution to the thin-film solar cells. This review puts light on the low efficiency of the metal sulfides and thus suffers the device's performance [24].

4.3.2 Ternary Thin-Film Solar Cell

Using a 1-D solar cells simulator named Solar Cell Capacitance Simulator, the effect of temperature on CIGS has been studied experimentally. The study was done to test the dependency of characteristics of CIGS on temperature ranging from 25°C to 70°C. Two components open-circuit voltage (V_o) and maximum power conversion efficiency of cell report degradation [25]. The three-stage co-evaporation process has been utilized with the involvement of various Se-fluxes to produce $CuGaSe_2$ (CGS) thin films. Distinctive properties can be noticed during each growth stage which varies from the ones produced with constant Se-flux in all three stages. Ga_2Se_3 precursor viz. created during the first stage drives the parameters like the size of the grain size, producing through orientation and width of the depletion region in CGS's solar cell device. Se-flux during the second and third stages determines the spectra of photo-absorption and device parameters. All this takes place in the process where CGS film was formed near-surface region [26]. The Cu_2SnS_3 ternary compound based on the structure of CuSnS Kesterite thin-film solar cell is composed of the electro-deposition along with Cu_2ZnSnS_4 and SnS [27]. Earth-abundant and cost-efficient elements compose the p-type semiconductor Cu_2SnS_3 [28] and depict optoelectronic properties which are comparable w.r.t. Cu_2ZnSnS_4 and SnS, thus making it a suitable contender for future photovoltaic implementation. This work follows the annealing procedure with an electro-deposited precursor in a sulfur and tin sulfide environment the ternary compound has been produced via the X-ray diffraction process was used for obtaining the absorber layer which was then investigated structurally and the crystal structure obtained indicates it to be monoclinic. The calculations of its properties based on optical parameters have been attained through photoluminescence. Investigation of CdMnTe thin films is done by structural, morphological, compositional, electrical and optical properties with the use of scanning electron microscopy, X-ray diffraction, UV is spectroscopy, sputtered neutral-mass spectroscopy and photo electrochemical cell measurements, respectively [29].

4.3.3 Quaternary Thin-Film Solar Cell

Two techniques were utilized to prepare Cu_2ZnSnS_4 (CZTS) absorber films: one is three RF source-based co-sputtering and other is annealing process that was carried out in the sulfurized atmosphere. This has offered a conversion efficiency of about 6.7%. The second technique was based on co-evaporation. By the use of vacuum co-evaporation, CZTS films [30] were produced on Si (1 0 0) using elemental sources like Cu, Sn, S and binary ZnS. The X-ray diffraction patterns showed suppressed and oriented growth of polycrystalline induced in a film produced at higher temperatures [31]. $Cu_{(2)}ZnSn(S;Se)_{(4)}$/(CZTSSe) thin films are fabricated by a new solution-based method. Synthesis of binary & ternary chalcogenide nanoparticles was done, and CZTSSe thin films were formed using these precursors. During thermal processing the characterization data material revealed bilayer micro-structure's formation and also suggested a forward path on the improvement of device [32]. The TW ranged application is better handled by the chalcogenide PV absorbers compared to the CZTS/CZTSSe as they make use of abundant constituent elements. These chalcogenide compounds include both XZ2 (where X = Cu, Sn, Fe and Z = S or Se) and Cu_2MSnN_4 (where M = Fe, Mn, Ni, Ba, Co, Cd and N = S or Se) compounds [33]. Cu_2BaSnS_4 semiconductor is quoted as a wideband p-type quaternary chalcogenide. It is quoted as a propitious hole transporting compound in the form of inverted perovskite $CH_3NH_3PbI_3$ thin-film solar cells. The reason behind it is the satisfactory stability in chemical mode and the high number of carrier mobility (10 cm^2/V s & apt band arrangement with $CH_3NH_3PbI_3$ [34].

4.4 THEORETICAL THIN-FILM SOLAR CELLS

The basic theoretical approach to handle the different aspects of material can be mathematical models, *ab initio* calculations or computational simulations. When we talk about materials science, the notion originates basically as a structure-property paradigm: Atoms are the basic constituents at the microscopic level of materials, and their behavior is determined by the interactional aspects observed at the microscopic level (measured in nanometers (nm) and femtoseconds (fs)) and at the macroscopic scale (order of centimeters (cm), milliseconds (ms) and more) for technological applications. The proposed plan for performing simulations based on materials over the characteristic length and time ranges has played an eminent role based on the technological revolution.

The computational method follows a well-established and new interdisciplinary research technique termed 'Computational Physics' or 'Computational Materials Science'. Simulations under the material physics revolve around lattice and defect dynamics tests that are carried out at atomic scale which is derived from the solutions of non-relativistic Schrodinger equation. It is done for a defined number of atoms [35]. A useful tool in accelerating the discovery of materials (especially optoelectronic semiconductor fields) is the high-throughput computational materials screening. Material databases are the place where we can find functional materials with a wide range of varied aspects of physical properties and applications [36]. Various computational tools to compute the different properties of materials (e.g., electronic, structural, optical, magnetic, etc.) are Quantum Espresso, VASP, WIEN2k, etc. A brief explanation of the theoretical tools is given as follows:

i. **Quantum Espresso**: An integrated suite consisting of open-source computer codes is known as Quantum Espresso and is used for electronic and structured calculations. Here, the material modeling is accomplished at the nanoscale range. The basis of this computational tool is a density functional theory (DFT), pseudopotentials and plane waves. It is quoted as an ensemble of programs that can be used to calculate the properties [37].

ii. **Vienna ab initio simulation package (VASP)**: *Ab initio* quantum mechanical molecular dynamics which is based on DFT is performed through a tool/package called VASP. Here, the different pseudopotentials and the set of plane-wave basis are utilized. The approach involves local-density approximation implemented in VASP. It also involves the use of efficient diagonalization matrix schemes to perform an exact evaluation of instantaneous electronic ground state and efficient Pulay mixing. All previous problems that occurred in the Car–Parrinello method which consists of the integration of electronic and ionic motion equations are reduced in this. Using generalized gradient approximation (GGA), ultrasoft Vanderbilt pseudopotentials or the projector augmented wave method, the relation between the ions and the electrons is determined. The required amount of plane waves per atom is reduced in all such techniques for transition metals and foremost row elements, respectively. VASP is used to calculate forces and stress and used to ease up the atoms toward their instantaneous ground state [38].

iii. **Wien2k**: DFT-based Kohn Sham equation is solved using the augmented-plane-wave plus local orbitals (APW+lo) method deployed in the WIEN2k package. In the APW+lo method, all electrons contained in the core part and valence part are consistently present in a full-potential method and are implemented in the WIEN2k tool. Various categories of parallelization are present and varied improved numerical libraries are used along with solving properties computationally like electronic band structure or optimized atomic structure, nuclear magnetic resonance, electric polarization or shielding tensor. After a short and detailed description of the APW+lo method, they have made a detailed review of the utility, potential and attributes of WIEN2k [39].

4.4.1 BINARY THIN-FILM SOLAR CELL

Antimony sulfide (Sb_2S_3) thin films are quite interesting as they play the role of absorbing layer for solar cell applications. The first principle approach has been adopted to study the electronic and optical properties of Sb_2S_3 thin films. The DFT approach has been implemented in the WIEN2k package in the form of a highly accurate full-potential linearized augmented-plane-wave method. The optical properties analysis of the Sb_2S_3 thin films has shown better optical-based absorption lying in the spectrum of visible light and UV wavelengths. It is predicted to be used as an imbibing layer in solar cells and optoelectronic-based devices [40]. Cu_3N film's comprehensive study using first principles through the theory of density functional based on electronic, structural and optical properties was performed. From the first principles along with Hubbard term (LDA + U), the bandgap is calculated and it turns out to be indirect in nature with a value of about 1.4 eV for the Cu_3N film. This value lies nearer to the calculated experimental value, i.e., 1.44 eV which is achieved using UV–vis absorption spectrum [41].

4.4.2 Ternary Thin-Film Solar Cell

Thin-film growth and the *ab initio*-based calculations for ternary alloys of ZrTaN material were done along with their structural, mechanical and phase stability feature investigation. $Zr_{(1-x)}Ta_{(x)}N$ films have got their DFT investigation done within the generalized gradient approximation. Calculating electronic structure along with predicted lattice parameters, mechanical properties (like single-crystal elastic constants, polycrystalline elastic moduli) of ternary cubic rocksalt-structured $Zr_{(1-x)}Ta_{(x)}N$ compounds with structure [42]. Electronic, structural and optical properties were reported to be investigated for conducting ternary transition metal nitrides that consist of metals belonging to varied periodic table groups. Optical spectroscopy and first-principles calculations were used to study bonding, electronic, structural and optical properties of the growth of conducting $Ti_{(x)}Ta_{(1-x)}N$ film [43]. The structural parameters of $CuBr_{(x)}I_{(1-x)}$ ternary alloy compound, within the DFT parameters, have been appropriately attained through computational techniques. The range taken up for the dopant concentration in compound $CuBr_{(x)}I_{(1-x)}$ has been taken from 0 to 1 [44]. Semi-automated DFT measurements were done to investigate structural, mechanical properties and phase stability of transition metal di-borides. The early transition metal diborides such as VB^2 and TiB^2 seem to be chemically substantial with the AlB^2 structure type. But earlier transition metal diborides such as WB^2 and ReB^2 exist in the $W2B^{5-x}$ structure type which is more stable [44]. DFT calculation of $CuInSe^2$ film with doping of aluminum has been presented in a thin-film-based work with a vacuum size of 10 Å. It is observed to be offering a bandgap close to that of pure $CuInSe^2$ and thus states its apt layout. Analysis of electronic and optical features for this compound has been deduced by Trans-Blaha Modified Becke–Johnson functional with the bandgap of 1.02 eV.

With the increment in the amount of doping element Al the increment in the bandgap owes to an apt range of the visible spectrum. So it has defined the application in the usage of a thin film in optoelectronics field [45]. Most accurate results in terms of investigating structural, optical and electronic properties are provided by Tran-Blaha Modified Becke–Johnson potential approximation. A supercell of $2 \times 2 \times 2$ dimension is constructed for the thin-film $CuGaSe_2$ with an insertion of

10 Å vacuum. The doping percentage is kept as 11.11% in which specifically Al doping has been done at the gallium site to observe the effect on its bandgap. Energy band structure and density of states both come under the electronic properties category. Pure CuGaSe$_2$ offers a 1.17 eV energy bandgap and with doping the increment is about 1.27 eV [46].

4.4.3 Quaternary Thin-Film Solar Cell

The structural, optical and electronic properties of quaternary alloy Cd$_x$Zn$_{1-x}$Se$_y$Te$_{1-y}$ which is quoted to be technologically important have been calculated utilizing the DFT-based full-potential-linearized augmented-plane-wave method. Perdew–Burke–Ernzerhof generalized gradient approximation, Modified Becke–Johnson and Engel–Vosko generalized gradient approximation methods are used as exchange correlation potentials in computing structural and optoelectronic properties [47]. The nature of the bandgap observed in this kind of alloy is direct. Concluding this work states that ZnTe, GaSb, InAs and InP are apt substrates used in the growth of several zinc blende Cd$_x$Zn$_{1-x}$Se$_y$Te$_{1-y}$ quaternary alloys. For the first time, thin films of Cu$_2$FeGeS$_4$ have been acquired using the spray ultrasonic deposition technique by the use of Cu-Fe-Ge precursors. The process is followed in aqueous salt which is put over glass substrate at a temperature $T_s = 180°C$. It is followed by sulfurization in the presence of an inert atmosphere, i.e., 500°C-Ar-S8. Polycrystalline Cu$_2$FeGeS$_4$ showcasing a tetragonal structure is formed using the XRD and Raman spectroscopy analysis. DFT-based theoretical calculations show a direct bandgap. The optical bandgap is ranged from 1.72 to 1.8 eV for both experimental and theoretical observations along with a high absorption coefficient (>10^4 cm^{-1}) in the visible spectra. The results computed are quoted to be apt materials for the photovoltaic field [48]. The design of two newly discovered I$_2$-II-IV-VI$_4$ quaternary semiconductors Cu$_2$ZnTiSe$_4$ and Cu$_2$ZnTiS$_4$, was created. Their study in terms of crystal and electronic structure was done using the first-principles electronic structure calculations of DFT. As per the crystal structures, bandgaps of Cu$_2$ZnTiSe$_4$ and Cu$_2$ZnTiS$_4$ it is confirmed to source out from fully occupied Cu-3d VB and unoccupied Ti-3d CB and kesterite structure attains the ground state. The calculations further depict those bandgaps of Cu$_2$ZnTiSe$_4$ and Cu$_2$ZnTiS$_4$ are comparable with bandgaps of Cu$_2$ZnTSe$_4$ and Cu$_2$ZnTS$_4$, but the absorption coefficient (α) is observed to be almost twice larger. Thus the compound qualifies for the candidature of solar cell absorber [49,50]. The electronic, electrical and optical properties, investigated based on DFT and Boltzmann transport theory for the Kesterite CZTS performed under varied strain conditions. Tensile strain effects and biaxial compressive activity have led to an altered bandgap for the particular structure as shown in the results obtained with varied values of strain. In addition, the movement of compression toward shorter wavelengths is observed due to the edge of absorption. Under the techniques of dilatation and compression, pure CZTS shows that increasing the amount of dilation improved the conductivity of the material. CZTS can be made as a suitable absorbent material by exploiting through the physical property, for improved task execution in solar cells [51].

4.5 BULK INTERMEDIATE BAND SOLAR CELL

An innovative approach to tackle the energy needs is through a photovoltaic device named intermediate band solar cell (IBSC) which offers a theoretical efficiency limit of 63.2%. It is potentially relied on for providing increased photo-generated current without degrading the value of output voltage. The concept in practice goes in this way: an intermediate band is necessarily synthesized with a discovery of novel semiconductor material that offers a band that is present at level three within the existing bandgap. The electronic population is scattered in all three zones/bands which split into quasi-Fermi levels, i.e., splitting at room temperature and also the non-radiative recombination rates should not exceed a defined limit. Currently, this feature can be achieved with the help of two approaches that are being actively investigated. The approaches are (i) quantum dot arrays self-assembled and (ii) semiconductors doped with an impurity. Samples that contain quantum dots usually play the role of IBSCs at low-valued temperature but face issues in reporting the properties succeed at room temperature. Materials that are impurity-doped show profitable optical features and are currently a growing field for researchers working on the reduction of their non-radiative recombination rates [52]. In a work, two possible semiconductor materials were discovered for IBSCs, and these materials were utilized as a used case, to identify obstacles in the realization of IBSCs. For $ZnO/IB-Cu_2O/Cu_2O$ cell, the analysis of an IBSC system has been given with the major objective of analyzing and optimizing the intermediate band solar cell. This was done to judge whether Cu_2O and ZnO are promising IBSC materials [53]. Although intermediate bands are produced artificially which needs extra efforts if we can have natural intermediate bandgaps, these can also prove beneficial for research eliminating the doping activity. One such work is shown in which bandgaps of quaternary chalcogenide semiconductors are grouped as I_2-II-IV-VI_4 (Cu_2ZnSnS_4 and $Ag_2ZnSnSe_4$) had been mentioned as a promising light-absorbing semiconductor material. $Ag_2ZnSnSe_4$ offered a bandgap approximately nearer to the optimal bandgap and exists in the form of a wurtzite-kesterite structure based on the Luque and Martinez model. Thus, can be claimed as an ideal light-absorbing semiconductor material of the intermediate band solar cells category. The width of $Ag_2ZnSnSe_4$'s intermediate band is quite large, limiting its efficiency and with a suitable doping amount the cell and intermediate band solar cells [59] needed bandgap can be attained to shift Fermi energy toward the intermediate band [54]. The structural and optoelectronic features of N-, P-, As- and Sb-doped $Cu_2ZnSiSe_4$ alloys are investigated in one of the works based on DFT using cell and intermediate band solar cells [59] needed bandgap can be attained to shift Fermi energy toward the intermediate band [54].

The structural and optoelectronic features of N-, P-, As- and Sb-doped $Cu_2ZnSiSe_4$ alloys are investigated in one of the works based on DFT using a hybrid functional for calculations [55]. Here 's' states in the orbitals of the doped V-A group atom; p states of Se atoms and little amount of d states of 8 Cu atoms are responsible for the effect of doping-induced IB. The features like absorption coefficient experience an enhanced effect due to the presence of IB with two extra peaks of absorption lying in the range of visible light. A typical transmittance spectrum is formed due to the

Intermediate Bandgap Solar Cell Materials

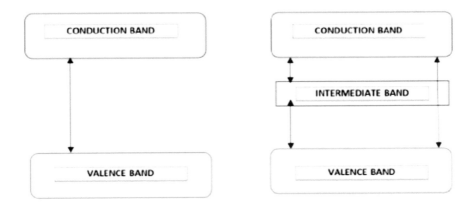

FIGURE 4.5 Schematic representation of absorption in a single-junction solar.

doped of Mn in GaN layers with a unique dip of around 820 nm caused as a result of the electron transition from the edge of the valence band to Mn-related states present within the bandgap. Doping of Mn in GaN in the bipolar devices results in the formation of electroluminescence spectra. It also shows that the energy states in the bandgap are present. Electroluminescence spectra are used to evaluate the location of different Mn-related energy states in GaN. In addition to this we can also observe sub-bandgap absorption caused due to intermediate band and is an effect of the Mn-related energy states [56]. The value of N_I doping density effect based on the short circuit current (I_{sc}), open-circuit voltage (V_o) and fill factor in the case of IBSC is analyzed. Among the different parameters of IBSC it is observed that maximum generation and efficiency, which is particularly caused at the value of IB occupation probability, tends to be very small in comparison to common value (i.e., 0.5) [57]. In another work, InAs quantum dot technology has been practically implemented for IBSCs and has presented an outline for the underlying theory which is based on IBSCs. The confined energy of the different states of an electron in InAs quantum dots results in the formation of IB and also forms three unique quasi-Fermi levels due to the separation of IB from CB (due to true zero density of states) [58]. In Figure 4.5, we can see the schematic representation of the intermediate band (IB) solar cell.

4.6 THIN-FILM INTERMEDIATE BAND SOLAR CELL

The Sn-doped chalcopyrite in the form of nanostructured particles & polycrystalline thin film is synthesized using novel methodologies. Sn doping introduces narrow IB which are partially filled and located at 1.7 eV ($CuGaS_2$) and 0.8 eV ($CuInS_2$) lying over valence band maximum in forbidden energy gap. Additional absorption and emission spectra deployed by IBs are revealed by the diffused reflection spectra and also the photoluminescence spectra which are further tested through first-principles calculations. The intermediate band is responsible for enhancing the photovoltaics, photo-catalysis and photo-induced production of hydrogen for wide solar spectrum response [60]. The evaluation of TF-IBSCs' potential of chalcopyrite absorbers has been done [61]. Bandgap engineering of material system's like $Cu(In; Ga)(S; Se)_2$ is performed by reforming contents of varied elements in groups III and VI, ranging from 0.95 eV in $CuInSe_2$ up to 2.4 eV in $CuGaS_2$. The addition of Al expands the energy gap further above 3.0 eV. A theoretical limiting efficiency for IB-TFSC has been reported for germanium monosulfide (GeS), and this material is observed to display wide bandgap (~1.7 eV), non-toxic, earth-abundant components and highly substantial features [62]. The high-buffer layer has been introduced among GeS and substrate that helps in achieving high-quality polycrystalline GeS thin film which follows to form a substrate with no particular gaps. Solar cells constructed based on GeS films depict a power conversion efficiency of 1.36% considered under AM 1.5G illumination [100 mW/cm^2]. The first-principles study of high contents of Cu and Al in thin-filmed $CuAlS_2$ is accomplished in which Sn and Fe doping in $CuAlS_2$ film has led to the formation of IB. Theoretical efficiency for Sn and Fe doped film is computed to be 34.67% and 35.64%, respectively. Sn-doped film's IB acts as a trapping location for the charge carriers. Thus it can be stated that Sn doping in $CuAlS_2$ film acts as a propitious IBSC compound [63]. An investigation has been carried out for the simulation of ordinary and extraordinary dielectric function components in different fractions of phase volume. To achieve this DF of wurtzite phase in ZnS and zinc blende phase in ZnS were utilized. ZnS films doped with Ti, Cu and In were deposited through thermal evaporation. Substitution doping was used to utilize the energy-dispersive spectroscopy at low doping levels. Degradation of the optical properties was observed for strongly doped samples in particular the Cu-doped samples showcased the same [64]. The Chromium doped $CuGaS_2$ Thin Films has been synthesized by chemical Spray Pyrolysis [65]. The addition of impurity or intermediate band in a semiconductor helps in enhancing optical functionalities of material in novel applications like IBSCs. The optical and electronic depiction for this purpose was done with chromium (Cr)-doped (1%–4%) chalcopyrite CGS thin film. *In situ* CGS thin films doped with Ti were layered by sputtering $CuGaS_2$: Ti targets to get IBSCs. Preparation of targets was done by use of hot-pressed sintering ball-milled Cu_2S, Ga_2S_3 and TiS_2 mixed powders. After sulfurization, grains grow and CGS phases are obtained [66]. The efficiency comparison for the varied kinds of thin films and intermediate band thin-film solar cells has been given in Table 4.2.

TABLE 4.2
Efficiency Comparison of Different Solar Cell

S. No.	Compounds	Efficiency
1	Bulk solar cell	(6%–7%) [67] 25% 20% [68]
2	Thin-film solar cells	10% [67]
3	Bulk intermediate solar cells	63.2% [49]
4	Thin-film intermediate solar cells	46.7% [69]

4.7 CONCLUSION

Finally, we would like to conclude our review with a brief description of how solar cells have emerged since their origin. The different classes of solar cells, their formation processes, their limitations and the solutions provided or discovered for the betterment of the field of energy have been discussed. Various categories of solar cells are explained with their binary, ternary and quaternary form depending on the number of elements in the sample. Two basic methodologies of the processing of these solar materials, i.e., experimental and theoretical approaches, have been discussed in this review. This writing is an overview review study based on the novel discovery of intermediate bandgap solar cells and their benefits have been briefly included. This new approach has proved to be very essential in improvising electricity generation. The efficiency of each of the solar cell compounds has been compiled to state the usefulness of that particular material compound category.

REFERENCES

1. York, R. and Bell, E.B., "Energy transitions or additions? Why a transition from fossil fuels requires more than the growth of renewable energy" *Energy Res. Soc. Sci.*, 51, 40–43 (2019).
2. Khatibi A., "Generation and combination of the solar cells: A current model review" *Energy Sci. Eng.*, 1–18 (2019).
3. Khan, K., et al. "Review on opto-electronic response of emerging solar photovoltaic materials" In *Energy Systems and Nanotechnology: Advances in Sustainability Science and Technology*. Springer, Singapore, 79–97 (2021).
4. Gray, J.L., *The Physics of the Solar Cell, Handbook of Photovoltaic Science and Engineering*. John Wiley & Sons, Ltd, New York, USA (2011).
5. Ranabhat, K., et al. "An introduction to solar cell technology" *J. Appl. Eng. Sci.*, 14, 405–491 (2016).
6. Badawy, W.A., "A review on solar cells from Si-single crystals to porous materials and quantum dots" *J. Adv. Res.*, 6, 123–132 (2015).
7. Gaur, A., et al. "Investigation of bulk, doped and thin film solar cells: A review article" In Dharmendra Tripathi and R. K. Sharma (eds) *Energy Systems and Nanotechnology: Advances in Sustainability Science and Technology*, Springer, Singapore, 1–21 (2021).
8. King, R. R., et al. "Solar cell generations over 40% efficiency" *Prog. Photovolt: Res. Appl.*, 20, 801–815 (2012).
9. Lee, T.D. and Ebong, A.U., "A review of thin film solar cell technologies and challenges" *Ren. Sustain. Energy Rev.*, 70, 1286–1297 (2017).
10. Sharma, S., et al. "Solar cells: In research and applications -a review" *Mater. Sci. Appl.*, 6, 1145–1155 (2015).
11. Zhang, H., et al. "Solar energy conversion and utilization: Towards the emerging photoelectrochemical devices based on perovskite photovoltaics" *Chem. Eng. J.* 393, 124766 (2020).
12. Green, M.A., "Crystalline and thin-film silicon solar cells: State of the art and future potential", *Solar Energy*, 74(3), 181–192 (2003).
13. Aberle, A.G., "Thin-Film solar cells" *Thin Sol. Films* 517, 4706–4710 (2009).
14. Lechner, P. and Schade, H. "Photovoltaic thin-film technology based on hydrogenated amorphous silicon" *Prog. Photovolt.* 10, 85–97 (2002).
15. Shockley, W. and Queisser, H.J. "Detailed balance limit of efficiency of p-n junction solar cells" *J. Appl. Phys.* 32, 510 (1961).

16. Plaza, A., et al. "CdTe thin films: Deposition techniques and applications" *Intech Open*, 1–18 (2018).
17. Gloeckler, M., et al. "CdTe solar cells at the threshold to 20% efficiency" *IEEE J. Photovolt.* 3, 1389 (2013).
18. Powalla, M., et al. "Thin-film solar cells exceeding 22% solar cell efficiency: An overview on CdTe⁻, Cu(In, Ga)Se₂₋, and perovskite-based materials" *Appl. Phys. Rev.* 5, 041602 (2018).
19. Ruiz, C.M., et al., "Experimental and simulation tools for thin film solar cells" (2016).
20. Dubey, R.S., et al. "Investigation of solar cell performance using multilayer thin film structure (SiO_2/Si_3N_4) and grating" *Results Phys.* 7, 77–81 (2017).
21. Montiel, K.A., et al. "Lead-free Perovskite thin film solar cells from binary sources" *IEEE 46th Photovoltaic Specialists Conference (PVSC)*, Chicago, IL, USA (2019).
22. Wang, Y., et al. "All-inorganic and lead-free BiI_3 thin film solar cells by iodization of BiSI thin films" *J. Mater. Chem. C* 8, 14066–14074 (2020).
23. Yeon, D.H., et al. "Origin of the enhanced photovoltaic characteristics of PbS thin film solar cells processed at near room temperature" *J. Mater. Chem. A* 2, 20112–20117 (2014).
24. Moon, D.G., "A review on binary metal sulfide heterojunction solar cells" *Solar Energy Mater. Solar Cells* 200, 109963 (2020).
25. Fathi, M., et al. "Study of thin film solar cells in high temperature condition" *Energy Proc.* 74, 1410–1417 (2015).
26. Ishizuka, S., et al. "Impact of a binary Ga_2Se_3 precursor on ternary $CuGaSe_2$ thin-film and solar cell device properties" *Appl. Phys. Lett.* 103, 269903 (2013).
27. Berg, D.M., et al. "Thin film solar cells based on the ternary compound Cu_2SnS_3" *Thin Sol. Films* 520(19), 6291–6294 (2012).
28. Chaudhari, J.J. and Joshi, U.S. "Fabrication of high quality Cu_2SnS_3 thin film solar cell with 1.12% power conversion efficiency obtain by low cost environment friendly sol-gel technique" *Mater. Res. Express* 5 036203 (2018).
29. Alam, A.E., et al. "Electrodeposition of ternary compounds for novel PV application and optimisation of electro-deposited CdMnTe thin-films" *Sci. Rep.* 10, 21445 (2020).
30. Katagiri, H., "Cu_2ZnSnS_4 thin film solar cells" *Thin Sol. Films* 480–481, 426–432 (2005).
31. Katagiri, H., et al. "Development of CZTS-based thin film solar cells" *Thin Sol. Films* 517, 2455–2460 (2009).
32. Cao, Y., et al. "High-efficiency solution-processed $Cu_2ZnSn(S, Se)_{(4)}$ thin-film solar cells prepared from binary and ternary nanoparticles" *J. Am. Chem. Soc.* 134, 15644–15647 (2012).
33. https://www.frontiersin.org/articles/10.3389/fchem.2019.00297/full.
34. Ge, J., et al. "Cu-based quaternary chalcogenide Cu_2BaSnS_4 thin films acting as hole transport layers in inverted perovskite $CH_3NH_3PbI_3$ solar cells" *J. Mater. Chem. A* 5, 2920–2928 (2017).
35. Steinhauser, M.O. and Hiermaier, S., "A review of computational methods in materials science: Examples from shock-wave and polymer physics" *Int J Mol Sci.* 10, 5135–5216 (2009).
36. Luo, S., et al. "High-throughput computational materials screening and discovery of optoelectronic semiconductors" *Wiley Interdiscipl. Rev.: Comput. Mol. Sci.* 11, (2021).
37. Cataldo, S.D., http://lampx.tugraz.at/hadley/ss2/bands/dft/calculations/cataldo/MAIN.pdf (2019).
38. https://icme.hpc.msstate.edu/mediawiki/images/8/8d/IntroDeckElectStrucCalc July28-2010.pdf.
39. Blaha, P., et al. "WIEN2k: An APW+lo program for calculating the properties of solids" *J. Chem. Phys.* 152, 074101 (2020).

40. Suleiman, A.B., et al. "First-principles calculation of optoelectronic properties of antimony sulfides thin film" *Phys. Sci. Int. J.* (2019).
41. Mukopadhyay, A.K., et al. "Optical and electronic structural properties of Cu_3N thin films: A first-principles study (LDA+U)" *ACS Omega* 5(49), 31918–31924 (2020).
42. Abadias, G., et al. "Electronic structure and mechanical properties of ternary ZrTaN alloys studied by ab-initio calculations and thin-film growth experiments" *Phy. Rev. B* 9060, 144107–144171 (2014).
43. Matenoglou, G.M., et al. "Structure and electronic properties of conducting, ternary $Ti_xTa_{1-x}N$ films" *J. Appl. Phys.* 105, 103714 (2009).
44. Walia, S.S., et al. "Theoretical investigation of structural properties of ternary alloy $CuBr_xI_{1-x}$ through first principle method" *J. Phys.: Conf. Ser.* 1531, 012046 (2020).
45. Moraes, V., et al. "Ab initio inspired design of ternary boride thin films" *Sci. Rep.* 8, (2018).
46. Gaur, A., et al. "Optical and electronic analysis of Al doped $CuInSe_2$ thin film based flexible solar cells" *AIP Conf. Proc.* 2294, 030003 (2020).
47. Khan, K., et al. "Revealing the impact of aluminum doping on optoelectronic properties of $CuGaSe_2$ thin films flexible solar cells - A DFT study" *AIP Conf. Proc.* 2294, 030004 (2020).
48. Chanda, S., et al. "Calculations of the structural and optoelectronic properties of cubic $Cd_xZn_{1-x}Se_yTe_{1-y}$ semiconductor quaternary alloys using the DFT-based FP-LAPW approach" *J. Comput. Electron.* 1 (2019).
49. Beraich, M., et al. "Preparation and characterization of Cu_2FeGeS_4 thin film synthesized via spray ultrasonic method-DFT study" *Mater Lett.* 275, 128070 (2020).
50. Wang, X., et al. "Crystal structure and electronic structure of quaternary semiconductors $Cu_2ZnTiSe_4$ and Cu_2ZnTiS_4 for solar cell absorber" *J. Appl. Phys.* 112, 023701 (2012).
51. Kahlaoui, S., et al. "Strain effects on the electronic, optical and electrical properties of Cu_2ZnSnS_4: DFT study" *Heliyon* 6, (2020).
52. Antolin, E., et al. "Intermediate band solar cells" *Comp. Renew. Energy* 1, 619–639 (2012).
53. Thorstensen, A.E., "Analysis of an intermediate band solar cell system: Based on systems engineering principles" (2013).
54. Liu, Q., et al. "Natural intermediate band in I2–II–IV –V I4 quaternary chalcogenide semiconductors" *Sci. Rep.* 8, (2018).
55. Yoshida, K. and Okada, Y. "Device simulation of intermediate band solar cells" *12th International Conference on Numerical Simulation of Optoelectronic Devices (NUSOD) Semantic Scholar*, Shanghai, China (2012).
56. Jibran, M., et al. "Intermediate band solar cell materials through the doping of group-VA elements (N, P, As and Sb) in $Cu_2ZnSiSe_4$" *RSC Adv.* 9, 28234–28240 (2019).
57. Lee, M.L., et al. "GaN intermediate band solar cells with Mn-doped absorption layer" *Sci. Rep.* 8, 8641 (2018).
58. Sikder, U., et al. "Effects of doping of intermediate band region on intermediate band solar cell characteristics" *7th International Conference on Electrical & Computer Engineering (ICECE)*, Dhaka, Bangladesh (2012).
59. Yang, X., et al. "Improved efficiency of InAs/GaAs quantum dots solar cells by Si-doping" *Solar Energy Mater. Solar Cells* 113, 14–147 (2013).
60. Yang, C., et al. "Observation of an intermediate band in Sn-doped chalcopyrites with wide-spectrum solar response" *Sci. Rep.* 3, 1286 (2013).
61. Marron, D.F., et al. "Thin-film intermediate band chalcopyrite solar cells" *Thin Solid Films* 517, 2452–2454 (2009).
62. Feng, M., et al. "Interfacial strain engineering in wide-bandgap GeS thin films for photovoltaics" *J. Am. Chem. Soc.* 143, 9664–9671 (2021).

63. Gaur, A., et al. "Role of intermediate band and carrier mobility in Sn/Fe doped $CuAlS_2$ thin film for solar cell: An ab-initio study" *Solar Energy* 215, 144–150 (2021).
64. Hope, R.B., "Doped ZnS thin films for intermediate band solar cells - deposition and characterization" NTNU Open (2015): http://hdl.handle.net/11250/2615558.
65. Ahsan, N., et al. "Characterization of Cr doped $CuGaS_2$ thin films synthesized by chemical spray pyrolysis" *Mech., Mater. Sci. Eng.* 1–9 (2017).
66. Yaowei, W., et al. "Fabrication of in-situ Ti-doped $CuGaS_2$ thin films for intermediate band solar cell applications by sputtering with $CuGaS_2$: Ti targets" *Vacuum* 169, (2019).
67. Dixon, A.E. "16- photovoltaic energy conversion: Theory, present and future solar cells" *Solar Energy Conver.* II, 243–259 (1980).
68. https://www.energy.gov/eere/solar/crystalline-silicon-photovoltaics-research.
69. Marti, A., et al. "Evaluation of the efficiency potential of intermediate band solar cells based on thin-film chalcopyrite materials" *J. Appl. Phys.* 103, 073706 (2008).

5 Finite Volume Numerical Analysis of Diamond and Zinc Nanoparticles Performance in a Water-Based Trapezium Direct Absorber Solar Collector with Buoyancy Effects

Sireetorn Kuharat, O. Anwar Bég, Henry J. Leonard, Ali Kadir, Walid S. Jouri, Tasveer A. Bég, B. Vasu, J.C. Umavathi, and R.S.R. Gorla

CONTENTS

Notation 78
Greek 78
5.1 Introduction 78
5.2 Mathematical Model 82
5.3 Mesh Independence Test and Validation of Finite Volume Code 84
5.4 Results and Discussion 85
    5.4.1 Effects of Rayleigh Number and Inclination on the Streamlines and Isotherms (Diamond) 85
    5.4.2 Total Heat Flux 91
    5.4.3 Surface Heat Transfer Coefficient 92
    5.4.4 Average Nusselt Number 93
    5.4.5 Local Nusselt Number 93
    5.4.6 Effects of Volume Fraction on the Streamlines and Isotherms (Diamond and Zinc) 94
    5.4.7 Heat Transfer Coefficient 94
5.5 Conclusions 96
References 97

DOI: 10.1201/9781003258209-6

NOTATION

AR: aspect ratio of enclosure
Cp: specific heat capacity of the base fluid (J/kg.K)
C_{pnf}: nanofluid specific heat (J/kg.K)
g: gravitational acceleration (m/s^2)
h: convective heat transfer coefficient (W/m^{2}K)
k_f: base fluid thermal conductivity (W/mK)
k_{nf}: nanofluid thermal conductivity (W/mK)
k_S: nanoparticle thermal conductivity (W/mK)
L: height of the enclosure (m)
Nu: Nusselt number (–)
$q''_{w\ CFD}$: ANSYS FLUENT heat flux term (W/m^2)
Ra: Rayleigh number (–)
T: temperature (K)
T_H: temperature of hot wall (K)
T_C: temperature of cold wall (K)
V_f: volume of fluid (m^3)
V_{np}: nanoparticle volume (m^3)
x: co-ordinate parallel to enclosure base wall (m)
y: co-ordinate parallel to enclosure base wall (m)

GREEK

α_m: thermal diffusivity (m^2/s)
β: coefficient of thermal expansion, α_m is thermal diffusivity
ΔT: temperature difference between the hot and cold walls (i.e. $T_H - T_C$)
ϕ: volume fraction of nanoparticles in base fluid (water)
μ_{nf}: dynamic viscosity of nanofluid (kg/m.s)
μ_f: dynamic viscosity of base fluid (water) (kg/m.s)
ρ: density (kg/m^3)
ρ_{nf}: nanofluid density (kg/m^3)
ρ_f: base fluid density (water) (kg/m^3)
ρ_s: metallic nanoparticle density (kg/m^3)

5.1 INTRODUCTION

Nanomaterials are increasingly being used in modern engineering and medical applications. These materials are engineered at the nanoscale and include carbon nanotubes, nanoshells, nanoparticles, nanogels, nano-powders and nanowires. Nanofluids are a subset of nanomaterials which were introduced at Argonne Energy Labs, USA in the mid-1990s [1]. They are synthesized by suspending carbon-based (e.g. silica, diamond) or metallic nanoparticles (e.g. copper, tin, zinc, aluminum oxide, etc.) in base fluids (e.g. water, ethylene glycol, lubricants, etc.). The resulting mixture has been shown to achieve significantly higher thermal conductivities compared with the original base fluid [2,3]. Keblinski et al. [4] (among others) have investigated in detail the possible mechanisms which contribute to thermal enhancement in nanofluids. They have shown that with smaller nanoparticles generally thermal conductivity is increased and this may be attributable to a variety of causes including molecular-level layering of the liquid at the liquid/particle interface, Brownian motion of the particles, micro-convection heat transport in the nanoparticles, nanoparticle clustering, etc., and that ballistic collisions dominate diffusive contributions in the boost in heat transport. The exceptional properties of nanofluids have mobilized their ever-widening applications in many branches of technology in recent years. This has been accompanied by significant research activity in mathematical modeling and computational simulations to provide a greater understanding of nanoscale transport which can benefit the optimization of nanofluid deployment. Representative studies include Thumma et al. [5] (on magnetic nanofluid surface deposition), Bég et al. [6] (on enrobing flows of aerospace components with smart nano-coatings), Bég et al. [7] (on drilling fluid doping with titanium and silica nanoparticles for enhanced lubricity), Bég et al. [8] (on orthopedic smart tribology) and Ali et al. [9] (on medical nano-pharmacodynamics). Many different numerical techniques have been utilized to solve the generally nonlinear boundary value problems associated with such nanoscale flows. Vasu et al. [10] used the Buongiorno two-component nanoscale model and a homotopy analysis method to study the time-dependent non-Boussinesq nanofluid convective heat and mass transfer from a spinning sphere. Bég et al. [11] deployed MAPLE numerical quadrature and the Tiwari–Das nanoscale model to study the performance of titanium, silver, copper and aluminum oxide metallic nanoparticles in solar magnetic coating flows with electromagnetic induction effects. Shamshuddin et al. [12] used the Adomian decomposition method to compute the effects of thermal radiation, homogeneous–heterogeneous reactions on magnetized Sisko rheological nanofluid flow from a stretching sheet in a permeable medium. Prakash et al. [13] derived numerical shooting quadrature and finite element solutions for electro-kinetic nanofluid pumping in microchannel slip flow with hybrid nanoparticles. Hiremath et al. [14] used a Crank–Nicolson finite difference code to study the time-dependent elastic-viscous nanofluid axisymmetric natural convective boundary layer enrobing flow over a heated circular cylinder. Bhardwaj et al. [15] used Lie group algebraic methods and homotopy simulation to conduct a second law thermodynamic optimization analysis of nanofluid slip stagnation flow from a perforated stretching surface. Very recently Bhatti et al. [16] employed the successive linearization method to compute the effects of hydrodynamic wall slip and radiation heat transfer on magnetite nanofluid stretching flow with cross-diffusion effects.

An important application of nanofluids in energy engineering is the area of direct absorber solar collectors (DASCs) [17]. These feature a working fluid which traps solar radiation in an enclosure and circulates the heat for subsequent use in a variety of technologies including electricity generation, domestic heating, etc. To better predict the performance of such solar absorbers, mathematical and numerical simulation of enclosure convection flows has become an indispensable tool. Many excellent studies have appeared in the past three decades considering a diverse range of solar collector designs and working fluids. Such investigations generally feature Navier–Stokes viscous flow models with an energy equation including thermal buoyancy (and other) effects. Reindl et al. [18] studied thermal convection in insulated cylindrical enclosures as a model of solar thermal storage tanks with immersed coil heat exchangers. They solved the unsteady Navier–Stokes and energy equations with a finite element method for Rayleigh numbers from 10^3 to 10^6, delineating pure free convection and mixed convection regimes via scale analysis. Amber and O'Donovan [19] analyzed two-dimensional unsteady natural convection driven by the direct absorption of concentrated solar radiation by high-temperature molten salt-filled enclosures for height to diameter ratios of 0.5, 1 and 2 and Rayleigh numbers 10^7–10^{11}. They assumed the vertical boundaries to be rigid and adiabatic vertical walls, with a base heat-conducting wall of finite thickness and an open adiabatic top surface subjected to a non-uniform concentrated solar flux. They computed streamline and isotherm contours, identified a nonlinear temperature profile, and further showed that velocity distribution is strongly affected by elapse in time. Elshamy and Ozisik [20] studied the steady-state laminar natural convection in air bounded by a hot plate and a cold cylindrical enclosure, deriving a correlation for average Nusselt number over the range of Rayleigh number from 10^5 to 10^6 for different values of the width-aspect ratio and thickness-aspect ratio of the plate. They showed that increasing aspect ratio boots the average Nusselt number as does Rayleigh number and further identified a two-cell pattern at and below critical values of an aspect ratio of 1.5. Further studies include Hammami et al. [21] (who used TRNSYS software to study unsteady free convection in a square enclosure with mass transfer gradient effects), Noorshahi et al. [22] (who examined natural convective flow of air in a tilted enclosure), Das and Basak [23] (who employed a Galerkin finite element code to simulate solar convection in a square, triangular-type 1 and triangular-type 2 (inverted triangle) enclosures) and Trombe et al. [24] (who considered radiative convection with buoyancy effect sin compound rectangular solar enclosures).

An alternative geometry which has been increasingly deployed in commercial solar systems is the trapezium. This has been found to achieve easier implementation in orientation toward the sun and large-scale manufacturing [25]. Engineers have therefore explored the implementation of a variety of nanofluids in both two-dimensional trapezium and three-dimensional trapezoidal enclosures in recent years. Job et al. [26] employed a mixed finite element method with polynomial pressure projection stabilization to compute the transient magnetic free convection flows of alumina-water and single-walled carbon nanotube-water nanofluids within a symmetrical wavy trapezoidal enclosure. They considered the case wherein the wavy lower boundary is instantaneously elevated raised to a constant hot temperature with the upper boundary sustained as thermally insulated. Akbarzadeh and Fardi [27] utilized the finite volume method and the Patankar SIMPLER algorithm to investigate the free convection of nanofluids with variable properties inside two- and three-dimensional channels with trapezoidal cross-sections. They observed that with stronger thermal buoyancy (i.e. higher Rayleigh number) heat transfer rates are enhanced in both geometries. They further showed that Nusselt number is reduced with an increase in the nanoparticle volume fraction from zero to 2% whereas it is subsequently elevated for nanoparticle volume fraction greater than 2% in the two-dimensional case. They also showed that greater inclination of the channel trapezoidal cross-section walls boosts the heat transfer rate.

Whereas the majority of studies of solar nanofluids have considered metallic nanoparticles (copper, aluminum, etc.) or carbon nanotubes suspended in water base fluid, relatively few have addressed diamond-water nanofluids. Diamond, a natural allotrope of carbon has impressive thermal, mechanical, and electrical properties [28]. Nano-diamond-nanofluid systems have therefore also attracted interest very recently owing to their potential in achieving a stable and consistent increase in thermal conductivity [29]. Sani et al. [30] presented experimental results light-intensity dependent optical properties of graphite/nano-diamond suspensions in ethylene glycol for direct absorption solar collectors and solar vapor generation. Branson et al. [31] used de-aggregation of oxidized ultra-dispersed diamond in dimethylsulfoxide to study the thermal performance of nano-diamond–poly(glycidol) polymer brush: ethylene glycol nanofluids, observing that a 12% thermal conductivity enhancement is achieved with a 0.9 volume fraction. Many other laboratory-based studies on viscosity and thermal property modification achieved with diamond nanofluids have been reported including Mashali et al. [32], Kumar et al. [33], Chin et al. [34] and Sundar et al. [35]. Quite recently several computational analyses of diamond nanofluids have also appeared in the scientific literature. Izadi et al. [36] presented finite element solutions for natural convection of different nanofluids (water-copper, water-diamond, and water-silicon dioxide nanofluids) inside a porous medium annular cylindrical enclosure using the Buongiorno model was utilized to track the nanoparticles concentration. They observed that water-diamond achieves the highest heat transfer rates whereas water-silicon dioxide produces the lowest. Jang and Choi [37] investigated theoretically the cooling performance of a microchannel heat sink with 6 nm copper-in-water and 2 nm diamond-in-water nanofluids.

Another promising metallic material which can be deployed in nanofluids is zinc which has excellent anti-corrosion properties and is only a moderately reactive metal making it useful for solar collector nanofluids. It is also the fourth most abundant metal employed in industry after iron, aluminum and copper. It has been studied both in its pure form and oxide form for nanofluid deployment in solar energy systems. Important studies in this regard include Wang et al. [38] (zinc oxide oil-based nanofluids for DASCs), Radkar et al. [39] (ZnO water nanofluid in helical copper tube heat exchangers in which 18.6% elevation in Nusselt numbers was achieved for 0.25 vol% of ZnO nanoparticles), Ali et al. [40] (zinc oxide water nanofluid automobile radiators), Zhang et al. [41,42] (anti-bacterial and sterilization medical applications), Khatak et al. [43] (electronic cooling), Lee et al. [44] (thermal enhancement) and Yu et al. [45] (viscosity measurements of zinc-water/ethylene glycol nanofluids).

In the present study a computational fluid dynamics analysis of steady-state incompressible thermal convection in a two-dimensional trapezium solar collector geometry is conducted to evaluate the relative performance of both carbon-based (i.e. diamond) and metal-based (i.e. zinc) nanoparticles. The Tiwari–Das formulation is implemented to compute viscosity, thermal conductivity and heat capacity properties for diamond-water and zinc-water nanofluids at different volume fractions. A finite volume code (ANSYS FLUENT ver 19.1) [46] is deployed. Laminar Newtonian flow is examined. The SIMPLE solver is utilized, and residual iterations are utilized for convergence monitoring. Mesh independence is included. Verification with the penalty finite element computations of Natarajan et al. [47] for the case of a Newtonian viscous fluid (zero volume fraction) is also conducted and excellent correlation is achieved. Isotherm, streamline and local Nusselt number plots are presented for different volume fractions, sloping wall inclinations (both negative and positive slopes are considered) and Rayleigh numbers.

5.2 MATHEMATICAL MODEL

Computational fluid dynamics software (ANSYS FLUENT) [46] is used to simulate a two-dimensional model of natural convection flow of diamond and zinc-water-based nanofluids in a trapezium enclosure, as illustrated in Figure 5.1. The nanofluid (solar absorber liquid) flow is considered to be laminar, steady-state and incompressible. The Tiwari–Das nanoparticle volume fraction model is deployed [11] and described in due course.

The fundamental equations for steady viscous, incompressible laminar flow are the Navier–Stokes equations. Thermal convection is analyzed via the Fourier-conduction-based energy conservation. The simulation is two-dimensional, time-independent and executed in a Cartesian (x, y) co-ordinate system. These equations take the following form:

Continuity Equation:

$$\frac{\partial u}{\partial x} + \frac{\partial v}{\partial y} = 0 \tag{5.1}$$

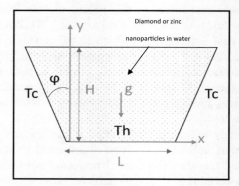

FIGURE 5.1 Trapezium water nanofluid direct absorber solar collector.

x-Direction momentum

$$u\frac{\partial u}{\partial x} + v\frac{\partial u}{\partial y} = -\frac{1}{\rho}\frac{\partial p}{\partial x} + v\nabla^2 u \qquad (5.2)$$

y-Direction momentum

$$u\frac{\partial v}{\partial x} + v\frac{\partial v}{\partial y} = -\frac{1}{\rho}\frac{\partial p}{\partial y} + v\nabla^2 v + g\beta(T - T_c) \qquad (5.3)$$

Energy equation

$$u\frac{\partial T}{\partial x} + v\frac{\partial T}{\partial y} = \frac{k_{nf}}{\rho_{nf} c_p}\nabla^2 T \qquad (5.4)$$

There is no heat transfer through the top wall (adiabatic condition). No-slip boundary conditions are assumed on all walls of the enclosure (cavity). Furthermore, the following thermal boundary conditions are imposed:

Left cold wall: Constant temperature, $T = 300$ K

Right cold wall: Constant temperature, $T = 300$ K (5.5)

Bottom hot wall: Constant temperature, $T = 305$ K

The remaining walls: Adiabatic

A Rayleigh number can also be defined as follows:

$$Ra = \frac{g\beta}{V\alpha m}(T_s - T_\infty)y^3 \qquad (5.6)$$

Rayleigh number (*Ra*) is a dimensionless number that signifies the ratio of thermal buoyancy and viscous force. Nusselt number (*Nu*) gives a measure for heat transfer rate along the hot wall of the two-dimensional enclosure. In Eq. (5.3), the term $(\rho\beta)_{nf} g(T-T_w)$ symbolizes the thermal buoyancy force. Since the changes in temperature and density of the nanofluid in this simulation are small, thus the Boussinesq model is employed. Boussinesq's model (named in honor of the great French engineer and mathematician J. Boussineq) considers density to be invariant for all the solved equations in ANSYS FLUENT [48]; this provides faster convergence in comparison to setting up the problem with fluid density as a function of temperature. This approximation is accurate provided $\beta(T-T_w)$ is less than or equal to unity. Absorber fluid, i.e. nanofluid transport is modeled as a "pseudo single-phase flow" since the nanoparticles are very small (the scale is one-billionth of a meter). Eqs. (5.1)–(5.4) with the associated boundary conditions (5) are solved by ANSYS FLUENT [47] software, and the temperature along with the solid–air interfaces is computed as part of the solution. This finite volume code has been deployed extensively by the authors in many areas including photovoltaic thermofluids [48], aerodynamics and gas dynamics [49–53] and nanofluid solar collector simulations [54–56]. The solution of the model gives the temperature and velocity fields and hence total heat flux across the heated wall, local and average Nusselt numbers along the heated wall, as well as the flow streamlines and surface shear stress. A nanofluid is defined in ANSYS FLUENT workbench as a "new fluid" with a new density, viscosity, thermal conductivity and specific heat obtained as a function of a base fluid and nanoparticle type and concentration (volume fraction), according to Brinkman as described in Bég [27]. The nanoparticle volume fraction can be estimated from:

$$\phi = \frac{v_{np}}{vf} \qquad (5.7)$$

The nanofluid dynamic viscosity can be estimated from:

$$\mu_{nf} = \frac{\mu f}{(1-\phi)^{2.5}} \qquad (5.8)$$

The effective nanofluid density and heat capacity also can be estimated from:

$$\rho_{nf} = (1-\phi)\rho f + \phi \rho s \qquad (5.9)$$

$$Cp_{nf} = \frac{(1-\phi)(\rho Cp)_f + \phi(\rho Cp)_s}{\rho_{nf}} \qquad (5.10)$$

The effective thermal conductivity of fluid can be determined by the Maxwell–Garnet relation which is adopted in the Tiwari–Das model and is elaborated in detail in [11]:

$$\frac{Knf}{kf} = \frac{ks + 2kf - 2\phi(kf - ks)}{ks + 2kf - \phi(kf - ks)} \qquad (5.11)$$

Here k_{nf} = nanofluid thermal conductivity, k_f = fluid thermal conductivity and k_s = nanoparticle thermal conductivity. All properties of the diamond and zinc-water nanofluids at different volume fractions are given in Table 5.1.

5.3 MESH INDEPENDENCE TEST AND VALIDATION OF FINITE VOLUME CODE

Extensive grid independence tests of the trapezium mesh were conducted for air inside trapezium geometry with $\varphi = 30$ and Ra 10^5 as shown in Figure 5.2a. Within the grid-dependent study, five different uniform grids system of 1,288, 2,030, 3,572, 4,472 and 5,244 are conducted over here. The result in an average Nusselt number of the fluid increases with a number of the elements until a small difference is observed between 4,472 elements and 5,244 elements which can be considered negligible which indicated that the simulation is convergent. The finalized mesh is shown in Figure 5.2b. 5,244 triangular elements were deployed.

TABLE 5.1
Nanofluid Properties

Diamond-Water-Based Nanofluid			Zinc-Water-Based Nanofluid		
Volume fraction	0.02		**Volume fraction**	0.02	
Density	1,047.358	kg/m^3	Density	1,119.158	kg/m^3
Thermal expansions (B)	0.000195972	1/K	Thermal expansion (B)	0.000187	1/K
Specific heat (Cpnf)	3,932.941613	J/kg.K	Specific heat (Cpnf)	3,698.221	J/kg.K
Thermal conductivity (Knf)	0.625009294	W/m.K	Thermal conductivity (Knf)	0.624828	W/m.K
Dynamic viscosity (Unf)	0.001135738	kg/m.s	Dynamic viscosity (Unf)	0.001136	kg/m.s
Pr	7.146759331		Pr	6.72218	
Volume fraction	0.04		**Volume fraction**	0.04	
Density	1,097.616	kg/m^3	Density	1,241.216	kg/m^3
Thermal expansion (B)	0.000183228	1/K	Thermal expansion (B)	0.000169	1/K
Specific heat (Cpnf)	3,709.41643	J/kg.K	Specific heat (Cpnf)	3,311.999	J/kg.K
Thermal conductivity (Knf)	0.636557081	W/m.K	Thermal conductivity (Knf)	0.636209	W/m.K
Dynamic viscosity (Unf)	0.001195818	kg/m.s	Dynamic viscosity (Unf)	0.001196	kg/m.s
Pr	6.968406511		Pr	6.225232	
Volume fraction	0.06		**Volume fraction**	0.06	
Density	1,147.874	kg/m^3	Density	1,363.274	kg/m^3
Thermal expansion (B)	0.0001716	1/K	Thermal expansion (B)	0.000154	1/K
Specific heat (Cpnf)	3,505.464699	J/kg.K	Specific heat (Cpnf)	2,994.936	J/kg.K
Thermal conductivity (Knf)	0.647669463	W/m.K	Thermal conductivity (Knf)	0.647167	W/m.K
Dynamic viscosity (Unf)	0.001260444	kg/m.s	Dynamic viscosity (Unf)	0.00126	kg/m.s
Pr	6.822065209		Pr	5.83304	

To verify the accuracy of the ANSYS FLUENT computations, a comparison has also been made for the purely Newtonian fluid case with the exact trapezium geometry and boundary conditions employed in the finite element simulations of Natarajan et al. [47]. The comparison of temperature contours (isotherms) is shown in Figure 5.3.

Excellent correlation is achieved and therefore confidence in the present ANSYS FLUENT code is justifiably high.

5.4 RESULTS AND DISCUSSION

Ansys fluent results are presented in Figures 5.4–5.15. We consider the influence of individual parameters in turn in the ensuing discussion.

5.4.1 Effects of Rayleigh Number and Inclination on the Streamlines and Isotherms (Diamond)

The progression of flow and thermal fields within the trapezoidal enclosure for $Ra = 10^3$, 10^4, 10^5, and 10^6 and tilt angle, $\emptyset = 30°, 20°, 10°, 0°$, and $-10°$ are presented

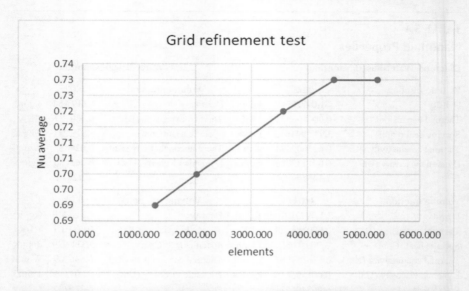

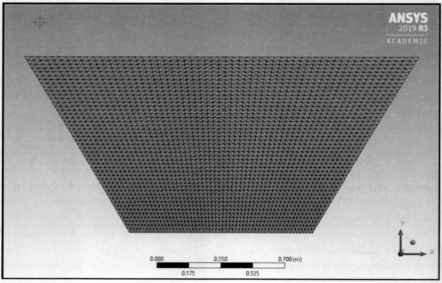

FIGURE 5.2 (a) Grid refinement test. (b) Final mesh design for nanofluid DASC.

in Figures 5.4–5.8. The flow and temperature fields are symmetrical about the vertical y-axis as the boundary conditions are the same for both left and right inclined sidewalls. As predicted, hot fluid mainly rises from the middle of the heated wall due to the strong effect of thermal buoyancy force (natural convection currents). The flow is redirected down along the walls forming two symmetric vortex cells with clockwise and anticlockwise rotations inside the cavity.

Finite Volume Numerical Analysis

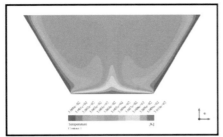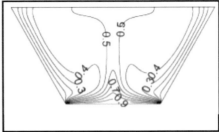

FIGURE 5.3 Comparison of ANSYS FLUENT and FEM solution of Natarajan et al. [47].

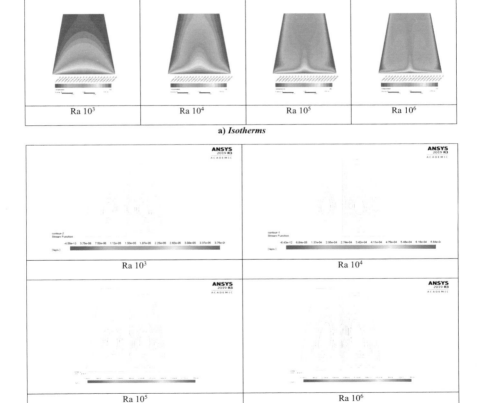

FIGURE 5.4 Diamond-water-based nanofluid with volume faction $\varphi = 0.02$, angle $(\emptyset) = -10°$.

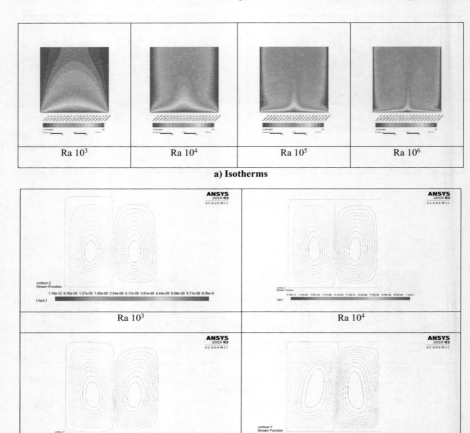

FIGURE 5.5 Diamond-water-based nanofluid with volume faction $\varphi = 0.02$, angle (Ø) = 0°.

With increasing Rayleigh number, the hot zone at the trapezium base is progressively suppressed. Heat transfer to the wall is therefore inhibited. However, there is a concurrent expansion in warmer fluid through the core region and the cooler zones at the sidewalls are contracted. Heat is therefore circulated more efficiently in the enclosure with an increasing thermal buoyancy effect. At a very high Rayleigh number (with volume fraction fixed at 2%, diamond nanofluid case; Figure 5.4), the streamlines are also morphed considerably, and the dual vortex structure is stretched in the vertical direction. The cells are increasingly lop-sided as compared with lower Rayleigh numbers. For the rectangular case (slope angle = 0) and the same volume fraction (2%, diamond case), the green zones are significantly expanded in the upper section of the enclosure with increasing Rayleigh number. Also, there is an elimination in the skewing of the dual vortex zones at higher Rayleigh numbers; however, some distortion is induced at the upper and lower wall zones in the streamlines which are not as controlled as in the trapezium case. With positive sidewall inclination (10°), as observed in Figure 5.6, there is a much more consistent distribution of heat achieved throughout the trapezium and this is encouraged with a greater Rayleigh number. However, at the highest Rayleigh number, colder regions emerge in a symmetric fashion which implies that a critical Rayleigh number between 106 and 107 exists where the best temperature distribution is obtained in the solar collector. The dual vortex structure witnessed in the streamlines is much more stable in Figure 5.6 than in previous figures (negative slope case, Figure 5.4; rectangular case, Figure 5.5). This would indicate that the strong positive slope trapezium produces the most regulated circulation compared with the other configurations. However, it does not achieve the best heat transfer to the boundaries, which is generally obtained with the strong

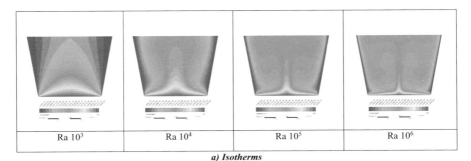

a) Isotherms

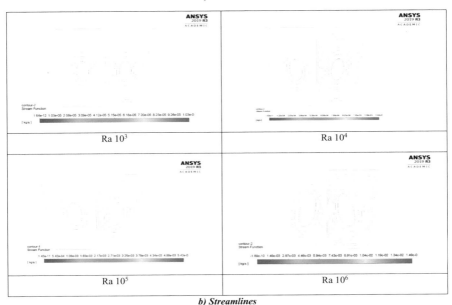

b) Streamlines

FIGURE 5.6 Diamond-water-based nanofluid with volume faction $\varphi = 0.02$, angle $(\emptyset) = -10°$.

negative slope case (Figure 5.4; $-10°$ inclination). While the trend in isotherms is sustained with an even greater positive sidewall slope (Figure 5.7), i.e. expansion in homogenous heat distribution, the vortex cells (streamlines) are increasingly warped with higher Rayleigh numbers, in particular at the top left and right corners of the cavity. A counterproductive distribution in both isotherms and streamlines is further witnessed with an even greater sidewall slope angle ($\emptyset = 30°$, i.e. gentle slope) in Figure 5.8. It would appear therefore that optimum results in terms of heat transmission to the boundaries are produced for the strong negative slope ($-10°$; Figure 5.4) even at relatively low diamond nanoparticle volume fraction (2%). It is also noteworthy that even at low Rayleigh numbers, e.g. $Ra = 10^3$, viscous forces are still significant and considerably greater than the buoyancy forces and, therefore heat transfer is essentially diffusion dominated and the shape of the streamline tends to follow the geometry of the enclosure. For $Ra = 10^5$ and 10^6, it is remarkable to observe that the temperature contours near the walls tend to

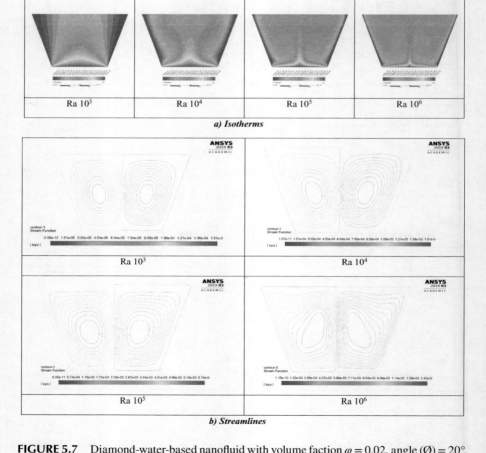

FIGURE 5.7 Diamond-water-based nanofluid with volume faction $\varphi = 0.02$, angle $(\emptyset) = 20°$.

Finite Volume Numerical Analysis

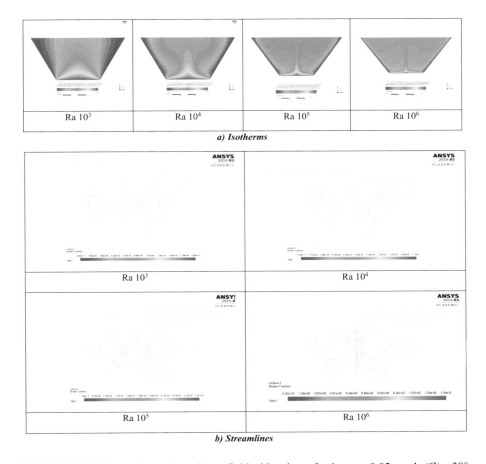

FIGURE 5.8 Diamond-water-based nanofluid with volume faction $\varphi = 0.02$, angle $(\emptyset) = 30°$.

have a neck formation caused by a strong circulation at a higher Rayleigh number which contrasts the flow pattern for $Ra = 10^3$. The core of the circulating rolls is narrowed (constricted) with an increase of Ra indicating a significant increase in the intensity of convection.

5.4.2 Total Heat Flux

Figure 5.9 shows the result for total heat flux with change in Rayleigh number for diamond-water-based nanofluid with volume fraction, $\varphi = 0.02$ at different sidewall inclination angles (30, 20, 10, 0 and −10). For a constant surface temperature, an increase in the total heat flux indicates a boost in the heat transfer to the walls. This corresponds to a decrease in temperature contour magnitudes within the DASC enclosure.

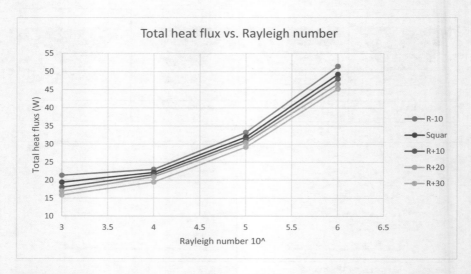

FIGURE 5.9 Total heat flux vs. Rayleigh number for diamond-water-based nanofluid with volume fraction $\varphi = 0.02$.

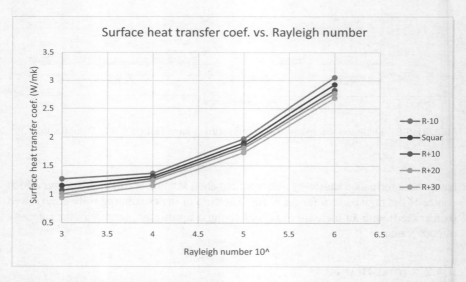

FIGURE 5.10 Surface heat transfer coefficient vs. Rayleigh number for diamond-water-based nanofluid with volume fraction $\varphi = 0.02$.

5.4.3 Surface Heat Transfer Coefficient

Figure 5.10 shows the results for average surface heat transfer coefficient at various Rayleigh numbers and different sidewall inclination angles, again for diamond-water-based nanofluid with volume faction $\varphi = 0.02$. Again, as shown in Figure 5.9, the negative sidewall case (angle of −10°) achieves the best heat flux to the wall, whereas an intermediate performance is achieved with the positive strong slope case (+10°).

5.4.4 Average Nusselt Number

Figure 5.11 illustrates the solutions for the average Nusselt number along the heated wall with different Rayleigh numbers at different sidewall inclination angles, for diamond-water-based nanofluid with volume faction $\varphi = 0.02$. At lower Ra for all cases of the sidewall inclination angle (30, 20, 10, 0 and −10), the graphs retain a flat trend indicating only a slight increase in Nu; Nusselt number however thereafter rises rapidly when Rayleigh number attains 10^4. Additionally, the average Nusselt number is higher for a lower sidewall inclination angle. For this case therefore compared with other geometries, a superior heat transfer rate is attained at the solar enclosure boundaries, at all Rayleigh numbers. As the sidewall inclination angles decrease, the trapezium morphs into a rectangular geometry and then an obtuse trapezium with positive sloping walls. The constant temperature sidewalls in the case of negative slope (−10°) act as heat sinks which move closer to the heat source resulting in improved heat transfer to the walls of the cavity.

5.4.5 Local Nusselt Number

The trends observed in Figure 5.11 are confirmed also in Figure 5.12 wherein the strong negative trapezium sloped case (inclination = −10°) again attains the highest magnitudes of local Nusselt number at all locations (x values) along the heated wall. The local Nusselt number is a maximum at $x = 0$ and $x = 1$ corresponding to the extremities of the heated boundary.

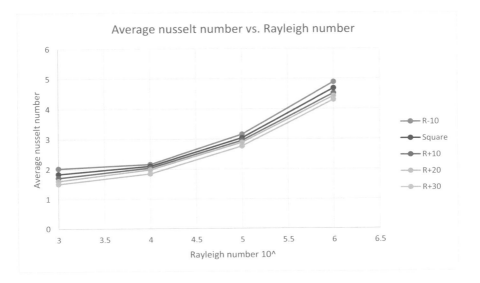

FIGURE 5.11 Average Nusselt number vs. Rayleigh number for diamond-water-based nanofluid with volume fraction $\varphi = 0.02$.

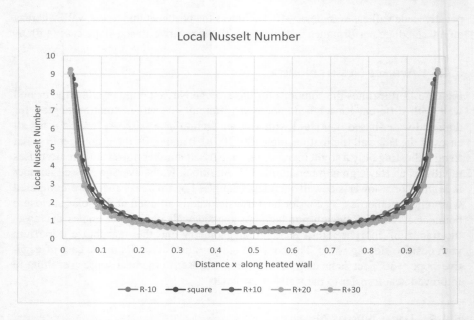

FIGURE 5.12 Local Nusselt number at the hot wall of enclosure for diamond-water-based nanofluid with volume fraction $\varphi = 0.02$.

5.4.6 Effects of Volume Fraction on the Streamlines and Isotherms (Diamond and Zinc)

Figures 5.13 and 5.14 illustrate the relative performance of diamond and zinc nanoparticles at different percentage doping, i.e. volume fractions, φ. The gentlest positive slope trapezium case of $\Phi = 30°$ is considered with the lowest Rayleigh number, $Ra = 10^3$. In both cases, there is a significant enhancement in temperature magnitudes with greater volume fraction. The hot zone at the base wall is expanded and warmer fluid (green/yellow zones) progressively replaces colder blue zones in the core region. Thermal diffusion in the regime is therefore clearly modified with greater quantities of nanoparticles. Slightly higher temperatures are computed for the zinc nanofluid compared with diamond nanofluid which is probably due to thermal conductivity differences. This inevitably leads to a difference in the heat transfer coefficient.

5.4.7 Heat Transfer Coefficient

Figure 5.15 depicts the variation in heat transfer coefficient versus volume fractions for both diamond and zinc-water nanofluids. At a low volume fraction (2%) both achieve approximately the same magnitude in heat transfer coefficient. However, with increasing volume fraction, the profiles diverge significantly and diamond-water

Finite Volume Numerical Analysis

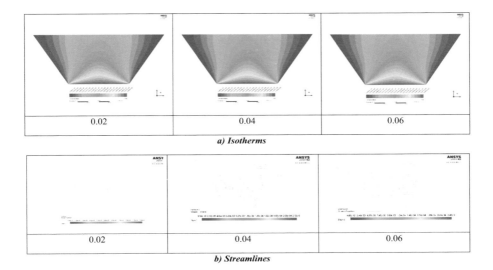

FIGURE 5.13 Diamond nanofluid in trapezium with 30° wall slope at Rayleigh number 10^3 with different volume fractions.

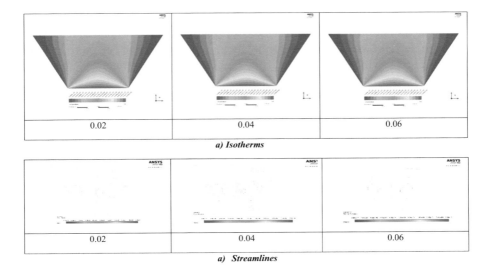

FIGURE 5.14 Zinc nanofluid in trapezium with 30° wall slope at Rayleigh number 10^3 at different volume fractions.

nanofluid produces a markedly higher heat transfer coefficient which is maximum at the highest volume fraction of 6%. Carbon nanofluids (diamond is an allotrope of carbon) are superior to metallic nanofluids in this regard.

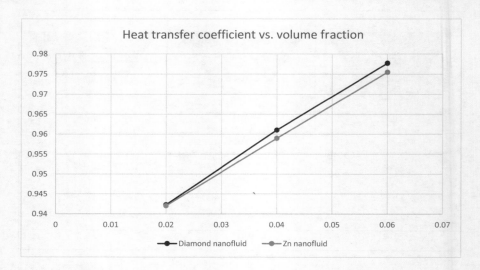

FIGURE 5.15 Heat transfer coefficient vs. nanoparticle volume fraction.

5.5 CONCLUSIONS

A theoretical and numerical study of the relative performance of both carbon-based (i.e. diamond) and metal-based (i.e. zinc) water-nanofluids in a trapezoidal geometry is presented. The Tiwari–Das formulation is implemented to compute viscosity, thermal conductivity and heat capacity properties for diamond-water and zinc-water nanofluids at different volume fractions. Steady-state nanofluid buoyancy-driven incompressible laminar Newtonian convection is examined. A finite volume code (ANSYS FLUENT version 19.1) is employed for the simulations with the SIMPLE solver, residual iterations are utilized for convergence monitoring. Mesh independence is included. Verification with the penalty finite element computations of Natarajan et al. [47] for the case of a Newtonian viscous fluid (zero volume fraction of any nanoparticles) is also conducted and excellent correlation is achieved. Isotherm, streamline and local Nusselt number plots are presented for different volume fractions, sloping wall inclinations (both negative and positive slopes are considered) and Rayleigh numbers. The present CFD simulations show that:

i. Vortex structure and thermal distributions are modified considerably with volume fraction, enclosure sidewall inclination and Rayleigh number.
ii. Diamond achieves higher heat transfer rates than zinc nanoparticles.
iii. Increasing heat transfer coefficients are computed with a rise in nanoparticle volume fraction for both diamond and zinc.
iv. Constant temperature sidewalls in the case of a strong negative slope ($-10°$) act as heat sinks which move closer to the heat source resulting in improved heat transfer to the walls of the cavity.

v. The strong positive slope trapezium case (inclination of +10°) produces the most regulated circulation when compared with the other configurations, although it does not achieve the best heat transfer to the boundaries – which is obtained with the strong negative slope case (i.e. −10° inclination).
vi. For the rectangular case (slope angle = 0) in the diamond-water nanofluid case, a more homogeneous temperature distribution is achieved throughout the enclosure compared with any trapezium scenario and the skewing of the dual vortex zones at higher Rayleigh numbers is eliminated.
vii. For the strong positive slope trapezium case (+10° inclination) compared with the strong negative slope trapezium case (−10° inclination), at very high Rayleigh number (volume fraction 2%, diamond nanofluid case) the streamlines are also morphed considerably, and the dual vortex structure is stretched in the vertical direction. The cells are increasingly lop-sided as compared with lower Rayleigh numbers for both sloped wall cases.

The present computations have furnished some interesting insights into the relative performance of zinc and diamond-water nanofluids in trapezium and rectangular solar enclosures. However, they have been confined to a steady-state and ignored porous media [57], three-dimensional [58, 59] and magnetohydrodynamic effects (magnetic nanoparticles). These are all of the great interest also in the next generation of "advanced functional material" solar direct absorber collectors and will be investigated in the near future.

REFERENCES

1. S. Choi, Enhancing thermal conductivity of fluids with nanoparticles, *Proceedings of the ASME International Mechanical Engineering Congress and Exposition*, San Francisco, CA. (1995).
2. S.K. Das, S.U.S. Choi, W. Yu and T. Pradet, *Nanofluids: Science and Technology*, Wiley, Hoboken, NJ. (2007).
3. X.Q. Wang and A.S. Mujumdar, Heat transfer characteristics of nanofluids: A review, *Int. J. Therm. Sci.*, Vol. 46, 1–19 (2007).
4. P. Keblinski, S.R. Phillpot, S.U.S. Choi and J.A. Eastman, Mechanisms of heat flow in suspensions of nanosized particles (nanofluids), *Int. J. Heat Mass Trans.*, Vol. 45, 855–863 (2002).
5. T. Thumma, O.A. Bég and A. Kadir, Numerical study of heat source/sink effects on dissipative magnetic nanofluid flow from a non-linear inclined stretching/shrinking sheet, *J. Mol. Liq.,* Vol. 232, 159–173 (2017).
6. O.A. Bég, A,S.. Rao, N. Nagendra, C.H. Amanulla, M. Surya, N. Reddy and A. Kadir, Numerical study of hydromagnetic non-Newtonian nanofluid transport phenomena from a horizontal cylinder with thermal slip: Aerospace nanomaterial enrobing simulation, *J. Nanofluids,* Vol. 7, 1–14 (2018).
7. O.A. Bég, D.E. Sanchez Espinoza, A. Sohail, T.A. Bég and A. Kadir, Experimental study of improved rheology and lubricity of drilling fluids enhanced with nanoparticles, *Appl. Nanosci.,* Vol. 8, 1069–1090 (2018).

8. O.A. Bég, A. Sohail, T.A. Bég and A. Kadir, B-spline collocation simulation of non-linear transient magnetic nano-bio-tribological squeeze film flow, *J. Mech. Med. Biol.*, Vol. 18, 1850007.1–1850007.20 20 pages (2018).
9. N. Ali, A. Zaman, M. Sajid, O. Anwar Bég, M.D. Shamshuddin and A. Kadir, Computational study of unsteady non-Newtonian blood flow containing nano-particles in a tapered overlapping stenosed artery with heat and mass transfer, *NanoSci. Technol.: An Inter. J.*, Vol. 9(3), 247–282 (2018).
10. B. Vasu, R.S.R. Gorla, O.A. Bég, P.V.S.N. Murthy, V.R. Prasad and A. Kadir, Unsteady flow of a nanofluid over a sphere with non-linear Boussinesq approximation, *AIAA J. Thermophys. Heat Transf.*, 13 pages (2018). Doi: 10.2514/1.T5516.
11. O.A. Bég, S. Kuharat, M. Ferdows, M. Das, A. Kadir and M. Shamshuddin, Magnetic nano-polymer flow with magnetic induction and nanoparticle solid volume fraction effects: Solar magnetic nano-polymer fabrication simulation, *Proc. IMechE-Part N: J Nanoeng, Nanomat. Nano-Syst.*, Doi: 10.1177/2397791419838714. 19 pages (2019).
12. M.D. Shamshuddin, S.R. Mishra, O.A. Bég and A. Kadir, Adomian computation of radiative-convective stretching flow of a magnetic non-Newtonian fluid in a porous medium with homogeneous–heterogeneous reactions, *Int. J Mod. Phys. B*, Vol. 33(2) 28 pages (2020).
13. J. Prakash, D. Tripathi and O.A. Bég, Comparative study of hybrid nanofluid performance in microchannel slip flow induced by electroosmosis and peristalsis, *Appl. Nanosci.*, 14 pages (2020). Doi: 10.1007/s13204-020-01286-1.
14. A. Hiremath, G. J. Reddy and O.A. Bég, Transient analysis of third-grade viscoelastic nanofluid flow external to a heated cylinder with buoyancy effects, *Arab. J. Sci. Eng.*, 19 pages (2019). Doi: 10.1007/s13369-019-03933-4.
15. P. Rana, N. Shukla, O.A. Bég and A. Bhardwaj, Lie group analysis of nanofluid slip flow with Stefan blowing effect via modified Buongiorno's model: Entropy generation analysis, *Diff. Eq. Dynam. Syst.*, 18 pages (2019). Doi: 10.1007/s12591-019-00456-0.
16. M.M. Bhatti, C.M. Khalique, T. Bég, O.A. Bég and A. Kadir, Numerical study of slip and radiative effects on magnetic Fe_3O_4-water-based nanofluid flow from a nonlinear stretching sheet in porous media with Soret and Dufour diffusion, *Mod. Phys. Lett. B*, 33, 2050026 24 pages (2020).
17. J.A. Duffie, W.A. Beckman and N. Blair, *Solar Engineering of Thermal Processes, Photovoltaics and Wind*, 5th Edition, Wiley, New Jersey, USA (2017).
18. D.T. Reindl et al., Transient natural convection in enclosures with application to solar thermal storage tanks, *ASME J. Sol. Energy Eng.*, Vol. 114(3), 175–181 (1992).
19. I. Amber and T.S. O'Donovan, Heat transfer in a molten salt filled enclosure absorbing concentrated solar radiation, *Int. J. Heat Mass Trans.*, Vol. 113, 444–455 (2017).
20. M.M. Elshamy and M.N. Ozisik, Numerical study of laminar natural convection from a plate to its cylindrical enclosure, *ASME J. Sol. Energy Eng.*, Vol. 113(3), 194–199 (1991).
21. M. Hammami et al., Transient natural convection in an enclosure with vertical solutal gradients, *Solar Energy*, Vol. 81, 476–487 (2007).
22. S. Noorishahi et al., Natural convection in a corrugated enclosure with mixed boundary conditions, *ASME, J. Sol. Energy Eng.*, Vol. 118(1), 50–57 (1996).
23. D. Das and T. Basak, Role of distributed/discrete solar heaters during natural convection in the square and triangular cavities: CFD and heatline simulations, *Solar Energy*, Vol. 135, 130–153 (2016).
24. Ar. Trombe et al., Solar radiation modelling in a complex enclosure, *Solar Energy*, Vol. 67, 297–307 (1999).
25. S. Kumar, Natural convective heat transfer in trapezoidal enclosure of box-type solar cooker, *Ren. Energy*, Vol. 29, 211–222 (2004).

26. V.M. Job et al., Time-dependent hydromagnetic free convection nanofluid flows within a wavy trapezoidal enclosure, *Appl. Thermal Eng.,* Vol. 115, 363–377 (2017).
27. P. Akbarzadeh and A.H. Fardi, Natural convection heat transfer in 2d and 3d trapezoidal enclosures filled with nanofluid, *J. Appl. Mech. Tech. Phys.,* Vol. 59, 292–302 (2018).
28. F. Mashali et al., Thermo-physical properties of diamond nanofluids: A review, *Int. J. Heat Mass Trans.,* Vol. 129, 1123–1135 (2019).
29. T. Tyler et al., Thermal transport properties of diamond-based nanofluids and nanocomposites, *Diamond Relat. Mater.,* Vol. 15, 2078–2081 (2006).
30. E. Sani et al., Graphite/diamond ethylene glycol-nanofluids for solar energy applications, *Renew. Energy,* Vol. 126, 692–698 (2018).
31. B.T. Branson et al., Nanodiamond nanofluids for enhanced thermal conductivity, *ACS Nano,* Vol. 7, 4, 3183–3189 (2013).
32. F. Mashali, E.M. Languri, J. Davidson and D. Kerns. Diamond nanofluids: Microstructural analysis and heat transfer study, *Heat Trans. Eng.,* Vol. 1, 1–13 (2020).
33. S. Kumar, M. Nehra, D. Kedia, N. Dilbaghi, K. Tankeshwar and K.-H. Kim. Nanodiamonds: Emerging face of future nanotechnology, *Carbon,* Vol. 143, 678–699 (2019).
34. R.-M. Chin, S.-J. Chang, C.-C. Li, C.-W. Chang and R.-H. Yu. Preparation of highly dispersed and concentrated aqueous suspensions of nanodiamonds using novel diblock dispersants, *J. Coll. Interf. Sci.,* Vol. 520, 119–126 (2018).
35. L. Syam Sundar, E. Venkata Ramana, M.P.F. Graça, M.K. Singh and A.C.M. Sousa. Nanodiamond-Fe_3O_4 nanofluids: Preparation and measurement of viscosity, electrical and thermal conductivities, *Int. Commun. Heat Mass Trans.,* Vol. 73, 62–74 (2016).
36. M. Izadi et al., Natural convection of a nanofluid between two eccentric cylinders saturated by porous material: Buongiorno's two phase model, *Int. J. Heat Mass Trans.,* Vol. 127, 67–75 (2018).
37. S.P. Jang and S.U.S. Choi, Cooling performance of a microchannel heat sink with nanofluids, *Appl. Thermal Eng.,* Vol. 26, 2457–2463 (2006).
38. X. Wang, Y. He, M. Chen and Y. Hu, ZnO–Au composite hierarchical particles dispersed oil-based nanofluids for direct absorption solar collectors, *Sol. Energy Mater. Sol. Cells.,* Vol. 179, 185–93 (2018).
39. R.N. Radkar et al., Intensified convective heat transfer using ZnO nanofluids in heat exchanger with helical coiled geometry at constant wall temperature, *Mater. Sci. Energy Technol.,* Vol. 2(7) 161–170 (2019).
40. H.M. Ali, H. Ali, H. Liaquat, H.T.B. Maqsood and M.A. Nadir, Experimental investigation of convective heat transfer augmentation for car radiator using ZnO-water nanofluids, *Energy,* Vol. 84, 317–324 (2015).
41. L. Zhang, Y. Jiang, Y. Ding, M. Povey and D. York, Investigation into the antibacterial behaviour of suspensions of ZnO nanoparticles (ZnO nanofluids), *J. Nanoparticle Res.,* Vol. 9, 479–489 (2007).
42. L. Zhang, Y. Ding, M. Povey and D. York, ZnO nanofluids – a potential antibacterial agent, *Prog. Nat. Sci.,* Vol. 18, 939–944 (2008).
43. P. Khatak, R. Jakhar and M. Kumar, Enhancement in cooling of electronic components by nanofluids, *J. Inst. Eng. India Ser. C,* Vol. 96, 245–251 (2015).
44. G.J. Lee, C.K. Kim, M.K. Lee, C.K. Rhee, S. Kim and C. Kim, Thermal conductivity enhancement of ZnO nanofluid using a one-step physical method, *Thermochim. Acta,* Vol. 542, 24–27 (2012).
45. W. Yu, H. Xie, L. Chen and Y. Li, Investigation of thermal conductivity and viscosity of ethylene glycol based ZnO nanofluid, *Thermochim. Acta,* Vol. 491, 92–96 (2009).
46. *ANSYS FLUENT Theory Manual, Version 19.1,* Lebanon, NH (2019).

47. E. Natarajan, T. Basak and S. Roy, Natural convection flows in a trapezoidal enclosure with uniform and non-uniform heating of bottom wall, *Int. J. Heat Mass Trans.*, Vol. 51, 747–756 (2008).
48. C.N.P. Sze, B.R. Hughes and O.A. Bég, Computational study of improving the efficiency of photovoltaic panels in the UAE, *ICFDT 2011-International Conference on Fluid Dynamics and Thermodynamics, Dubai, United Arab Emirates,* Dubai, January 25–27 (2011).
49. H.A. Daud, Q. Li, O.A. Bég and S.A.A. AbdulGhani, Numerical investigations of wall-bounded turbulence, *Proc. Inst. Mech. Eng.-Part C: J. Mech. Eng. Sci.*, Vol. 225, 1163–1174 (2011).
50. H.A. Daud, Q. Li, O.A. Bég and S.A.A. AbdulGhani, Numerical investigation of film cooling effectiveness and heat transfer along a flat plate, *Int. J. Appl. Math. Mech.*, Vol. 8(17), 17–33 (2012).
51. O.A. Bég, A. Zubair, S. Kuharat and M. Babaie, CFD simulation of turbulent convective heat transfer in rectangular mini-channels for rocket cooling applications, *ICHTFM 2018: 20th International Conference on Heat Transfer and Fluid Mechanics, WASET,* Istanbul, Turkey, August 16–17 (2018).
52. O.A. Bég, B. Islam, M.D. Shamsuddin and T.A. Bég, Computational fluid dynamics analysis of moisture ingress in aircraft structural composite materials, *Arab. J. Sci. Eng.,* 23 pages (2019). Doi: 10.1007/s13369-019-03917-4.
53. A. Kadir, O.A. Bég, M. E. El Gendy, T.A. Bég and M. Shamsuddin, Computational fluid dynamic and thermal stress analysis of coatings for high-temperature corrosion protection of aerospace gas turbine blades, *Heat Trans.*, 25 pages (2019). Doi: 10.1002/htj.21493.
54. S. Kuharat, O. Anwar Bég, A. Kadir and M. Babaie, Computational fluid dynamic simulation of a solar enclosure with radiative flux and different metallic nano-particles, *International Conference on Innovative Applied Energy (IAPE'19)*, St. Anne's College, Oxford University, Oxford, 14–15 March (2019).
55. S. Kuharat and O.A. Bég, Computational fluid dynamics simulation of a nanofluid-based annular solar collector with different metallic nanoparticles, *Heat Mass Trans. Res. J.*, Vol. 3(1), 1–23 (2019).
56. O.A. Bég, S. Kuharat, T.A. Bég, A. Kadir, H.J. Leonard and W. S. Jouri, Computation of radiative heat transfer in solar direct absorber collector flows with robust numerical methods and multi-physical effects. In *Understanding Thermal Radiation*, Ed. K.S. Rawat, Nova Science, New York, 200–310, (2020, September). In press.
57. B.M. Al-Srayyih et al., Natural convection flow of a hybrid nanofluid in a square enclosure partially filled with a porous medium using a thermal non-equilibrium model, *Phys. Fluids*, Vol. 31, 043609 (2019).
58. S. Kuharat, O.A. Bég, A. Kadir, B. Vasu, T.A. Bég and W.S. Jouri, Computation of gold-water nanofluid natural convection in a three-dimensional tilted prismatic solar enclosure with aspect ratio and volume fraction effects, *Nanosci. Technol.- An Int. J.,* Vol. 11(2), 141–167 (2020).
59. R.S.R. Gorla et al., Heat source/sink effects on a hybrid nanofluid-filled porous cavity, *AIAA J. Thermophys. Heat Trans.*, Vol. 31(4), 100–115 (2017).

6 Thermal Performance Study of a Copper U-Tube-based Evacuated Tube Solar Water Heater

Arun Uniyal and Yogesh K. Prajapati

CONTENTS

6.1 Introduction 101
6.2 Experimental Methodology 106
    6.2.1 Experimental Setup and Assembling of the Evacuated Tube Solar Collector 106
    6.2.2 Experimental Procedure 107
6.3 Thermal Performance Evaluation 108
    6.3.1 Incident Solar Radiation 108
    6.3.2 Daily Thermal Efficiency 108
    6.3.3 Uncertainty Analysis 108
6.4 Results and Discussion 109
    6.4.1 Temperature Distribution Inside the Collector 109
    6.4.2 Influence of Coolant Mass Flow Rate 110
    6.4.3 Thermal Performance Assessment 111
6.5 Conclusion 112
Nomenclature 112
References 113

6.1 INTRODUCTION

Rapid modernization, upgradation of living standards, and increasing population of the world have considerably increased the energy requirement. Furthermore, dependencies on people's accomplishments are being immensely overseen by energy-driven appliances. Development of any nation is benchmarked by the per capita energy consumption. All these circumstances have further accelerated the energy demand globally. Nevertheless, even today, the major source of energy is fossil fuels, which contribute $\approx 80\%$ to the total energy requirements [1,2]. Extraction of energy from fossil fuels causes environmental pollution, climate change, global warming, and several other harmful effects. Also, conventional sources of energy are not sustainable for a longer period. Therefore, in the current scenario, greater attention is given to renewable and sustained sources of energy. Subsequently, numerous forms of renewable energy sources such as solar, wind, ocean, biomass, and geothermal have been explored in the recent past. Among these renewable sources, solar energy is one of the most economic and available in abundance. Therefore, more emphasis has been given to optimally utilize solar energy, and hence greater research and development need to be done in this domain. Currently, the application of solar energy spreads over various fields, such as space heating, water heating, cooking, drying, electricity or power generation, desalination. Out of these applications, solar water heating is one of the most prevalent and environment-friendly methods to utilize solar energy.

Commonly, the following two types of solar water heaters are in use these days:

i. Flat plate solar water heater
ii. Evacuated tube solar water heater

It is worth mentioning that flat plate solar water heaters have been used for several years. They are simple in construction, having less initial cost with low maintenance. But, it has certain drawbacks such as poor thermal performance and a large amount of heat loss. Additionally, in such a type of design configuration, solar tracking is essentially required to utilize maximum solar radiation. However, an evacuated tube solar water heater is designed in such a way that it reduces most of the demerits of the flat plate heater. It consists of two annular glass tubes in which there is a vacuum in-between the annular space of the tubes. Such arrangement helps to minimize the heat loss from the heater. Furthermore, tubular design of the collector tube does not require any tracking of the solar radiation.

Evacuated tube solar collectors are also classified into three categories:

i. All water in glass tube evacuated solar collector
ii. U-tube-shaped evacuated solar collector
iii. Heat pipe-based evacuated solar collector

All water in glass tube evacuated solar collector consists of a set of glass tubes which are connected to a water storage tank. Vacuum is generated in the annular space in order to minimize the conduction and convection losses. The water flows from the storage tank to the glass tubes due to gravity. It works on the principle of thermosiphon. These tubes are the simplest and cheapest in configuration but applicable only to solar pre-heater or low-pressure water heaters [3].

U-tube-shaped evacuated solar collector consists of similar evacuated glass tubes; additionally, U-shaped copper tubes are inserted inside the hollow space available in the glass tube. Heat transfer fluid is being passed through the copper tube. These systems are particularly designed for the high-pressure and high-temperature-based applications. Their anti-freezing, easy installation, and quick start-up characteristics make these tubes more attractive compared to heat pipe-based solar evacuated tube solar water heaters. However, due to incorporation of copper tube and addition of extended fin surface results in higher initial cost and complex design of the system [4].

A heat pipe-based solar water heater consists of a set of tubes attached to a tank. Each tube consists of a heat pipe which is generally made up of copper. This heat pipe is surrounded by a glass tube and the annular space is usually evacuated. The heat pipe is a cylindrical container which consists of a capillary wick structure and heat transfer fluid is generally low melting point-based alcohol fluid. This heat pipe has three regions: (i) working fluid section, (ii) evaporative section, and (iii) condenser section that is working on evaporating–condensing cycle. The working fluid section is at the bottom of the heat pipe in which the working fluid is present. This working fluid takes the heat from the solar radiation and evaporates. It gives up its heat at the condenser section to the heat sink located at the top of the tube. Hence, heat transfer fluid flows by natural circulation between two phases in order to transfer the heat as needed [5].

Hence it may be concluded that U-tube is the most common arrangement in the evacuated tube solar collector due to its simplicity in design, practical feasibility, and having high thermal efficiency which makes it a cost-effective system. U-tube-type solar water heaters are most viable for high-pressure and high-temperature applications and have a more simple structure, compared to the all-glass tube and heat-pipe evacuated tube. Zhang et al. [6] have used supercritical CO_2 as a working fluid in a U-tube-type solar collector. The results showed that there is a significant impact of solar radiation on CO_2 temperature and pressure. The authors found daily time-weighted average collector efficiency to be more than 50%, and annual collector efficiency was reported nearly 60% which is higher than those conventional collectors using water as a working fluid. Ma et al. [7] analytically studied the thermal performance of a U-tube-type solar water heater using an energy balance equation in order to evaluate the heat loss coefficient and heat efficiency factor. The results showed that there is a nonlinear agreement between the heat loss coefficient of the evacuated tube solar collector with temperature gradient between the absorbing coating surface and the surrounding air. The heat efficiency factor was mainly affected by the ambient air layer between the absorber tube and copper fins. If synthetic conductance has been increased from 5 to 40 W/m K, solar collector efficiency and outlet fluid temperature are increased by 10% and 16%, respectively. Liang et al. [8] proposed a filled-type U-tube-based evacuated solar collector in which a high conductive material (compressed graphite) was used as a filler material between the U-tube and inner glass tube. Authors found that there was a 12% increment in thermal performance of a filled-type evacuated tube as compared to copper fin-type evacuated tubes when thermal conductivity reached up to 100. Liang et al. [4] presented a novel design of the evacuated tube collector (ETC) water heater in which a double U-tube was used inside the evacuated tube to eliminate the effects of thermal resistance between absorber plate and fins. It has been observed that thermal efficiency was 80% at the solar intensity of 900 W/m². Liang et al. [9] studied a universal model of a filled-type ETC. The authors have performed thermal performance of three different types of ETCs in which one evacuated tube is filled with a single U-tube, another tube is filled with two U-tubes and a third evacuated tube is filled with three U-tubes under the same boundary conditions. The results revealed that maximum solar collector efficiency was 82% for evacuated tube solar collectors equipped with three U-tubes, which is shown in Figures 6.1 and 6.2.

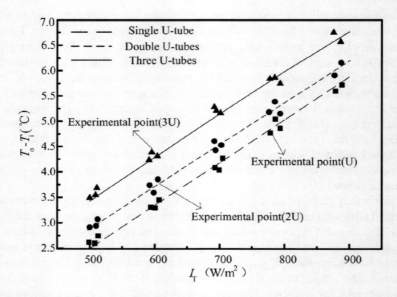

FIGURE 6.1 Variation of the temperature difference between outlet and inlet with the inclined solar irradiance [9].

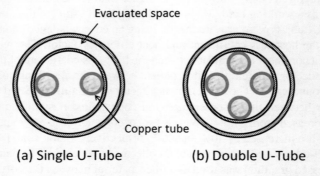

FIGURE 6.2 Representation of different configurations: (a) single U-tube and (b) double U-tube evacuated solar collectors [9].

Gao et al. [10] conducted a parametric analysis of a U-tube-type evacuated solar collector using mathematical modelling methods. The results showed that the thermal efficiency of the evacuated-type solar collector was not increasing with the length. They suggested that it will be better to use a low heat loss coefficient-based solar collector in cold regions. Pandey et al. [11] investigated the thermal performance of U-tube-based ETC using exergetic analysis. It was observed that the U-tube-based ETC system has maximum efficiency at 15 L/h and a minimum at 30 L/h. Energy efficiency was 66.57% and exergetic efficiency was 13.38% at a mass flow rate of 15 L/h. Naik et al. [12] performed mathematical modelling and performance analysis of U-tube-type solar collectors using different working mediums (water, air, and

LiCl-H_2O). The results revealed that to increase the outlet temperature, low amount of mass flow rate and optimized collector length is required. Further, they concluded that there is a significant impact of changing the ambient temperature on the net heat gain of solar collectors at high solar intensity days. Additionally, the authors pointed out that water has the maximum amount of heat absorption capacity as compared to other fluids used in their study. Nie et al. [13] investigated the thermal performance of a U-tube-type evacuated solar collector in which working fluid was operating at a lower temperature compared to the ambient temperature. Their findings revealed that thermal efficiency is improved as the reduced temperature is decreased and thermal efficiency is high above the low solar irradiance. Korres et al. [14] performed the optical and thermal performance of mini compound parabolic collector U-tube-type evacuated solar heaters. It has been found that there is insignificant impact on collector efficiency by changing the longitudinal incidence angle, whereas the transverse incidence angle improves the optical efficiency. Naik et al. [15] experimentally and numerically investigated the performance of a U-tube-type evacuated solar water heater. The results showed that solar intensity, mass flow rate of the fluid, and inlet temperature of the working fluid have a significant effect on useful heat absorbed by the working fluid moving inside the U-tube and the average overall efficiency was reported around 51%. It has been found that the average predicted energy efficiency is 43% and exergy efficiency was 41% for humid subtropical conditions. Korres et al. [16] conducted analysis on an array of four U-tube-type evacuated solar collectors integrated with mini compound parabolic collectors operating in the range of −80°C to 35°C. The results indicate that thermal efficiency decreases from the first module to the last module because of the higher inlet temperature which causes heavy heat losses. Bhowmik et al. [17] developed a multi-layer perception approach in order to predict the performance of U-tube-based ETC. The maximum efficiency was found to be 75% for the mass flow rate of 0.08 kg/s. Badar et al. [18] investigated the overall heat loss coefficient of the evacuated tube solar collector. It has been observed that the amount of overall heat loss coefficient is nearly 2–4 W/m^2-K. Kim et al. [19] investigated the thermal performance of a U-tube solar collector using different nanofluids such as MWCNT, Al_2O_3, CuO, SiO_2, and TiO_2. Multiwall carbon nanotube has shown a maximum efficiency of about 62.5% at 0.2% of volume. Tong et al. [20] studied the effect of thermal performance of U-tube solar collectors using multiwall carbon nanotube as a working fluid. Results demonstrated that there is a 5% increase in the efficiency of the solar collector using multiwall carbon nanotube. However, nanofluid with a concentration of 0.24 vol% shows the maximum heat transfer coefficient between the copper tube and the working fluid. Xie et al. [21] developed composite phase change material equipped with a tankless solar collector. Composite PCM (Phase Change Material) is made up of a combination of stearic acid and coconut shell charcoal. It has been observed that there is a 2.88 times increment in thermal conductivity of composite material as compared to stearic acid. Olifan et al. [22] proposed a spiral corrugated U-shaped copper tube of evacuated solar collector integration with phase change material. Paraffin wax is taken as a phase change material. Results revealed that there is a 21.55% increment in the collector efficiency of the collector as compared to the smooth tube. Kaya et al. [23] investigated the thermal performance of a U-tube evacuated solar collector using

ZnO/ethylene glycol-pure water nanofluids as a working medium. They have concluded that maximum collector efficiency is found to be 62.87 for 3% of volume at a mass flow rate of 0.045 kg/s. Abokersh et al. [24] studied the performance of modified U-tube-type ETCs integrated with paraffin wax. An extended surface with an area of 0.1251 m² has been integrated with the system. By regression analysis, it was found that mean annual efficiencies for simple forced solar water heater, water heater without fins and with fins are 40.5%, 71.8% and 85.7%, respectively.

Based on the above literature survey, it may be concluded that a U-tube-based evacuated solar collector has reasonably good potential to extract solar heat and transfer it to the working medium. However, consistent efforts are continuously required to further improve the heat transfer rate and overall efficiency. Moreover, its applicability and versatility need to be examined in the various regions and locations. Considering the above fact, in the present work, a parametric study has been done through experimental investigation on the U-tube-type evacuated solar collector at various solar radiation intensities. Results of the heat transfer rate through the fluid inlet and outlet temperature and coolant mass flow rate have been presented for the daytime 10:00 AM–2:00 PM at NIT Uttarakhand Srinagar (Garhwal), Uttarakhand.

6.2 EXPERIMENTAL METHODOLOGY

6.2.1 Experimental Setup and Assembling of the Evacuated Tube Solar Collector

Evacuated tubes consist of two concentric glass tubes which are fused, with one end closed and the other open. Vacuum is generated in the annular space in order to avoid heat losses. As shown in Figure 6.3, in the present study, sets of ETC tubes have been formed. Each set consists of three ETC tubes which are integrated with U-tubes made up of copper materials. Water has been used as a working medium. The dimension and material of the evacuated tubes and U-tubes are illustrated in Table 6.1.

FIGURE 6.3 Photograph of the experimental setup.

TABLE 6.1
Geometrical Configuration and Material of ETC and U-tube

	Geometrical Configuration of ETC and U-Tube	
1	Length of ETC tube (mm)	1,800
2	Outer diameter of ETC tube (mm)	58
3	Inner diameter of U-tube (mm)	7
4	Material of U-tube	Copper

6.2.2 EXPERIMENTAL PROCEDURE

It is worth mentioning that the experimental study has been carried out in the outdoor condition of the National Institute of Technology, Uttarakhand. Figure 6.4 shows the different components of a U-tube-type evacuated tube solar water heater. The experimental setup consists of ETC tubes, U-tube, storage tank, pyranometer, thermocouples, data acquisition system, and flow meter. The setup consists of three evacuated tubes in which a copper U-tube is fitted. The experimental setup has been faced towards the south direction at an angle of 32.447° since NIT Uttarakhand is situated at a latitude of 32.447° and longitude of 78.798°. Water has been supplied in the ETC and circulated through copper U-tubes by fluid pump and water leaving the collector is being collected in the storage tank.

The instruments used in the experiment are shown in Figure 6.4. The solar radiation intensity is measured with the help of pyranometer which is placed towards the south direction at the same angle of the experimental setup. Three thermocouples have been placed inside the evacuated collector in order to measure the average inner temperature of the collector. Two thermocouples are placed at the inlet and outlet section of fluid in order to measure the temperature of water at those respective locations. The following measurements have been done and data of the same is collected

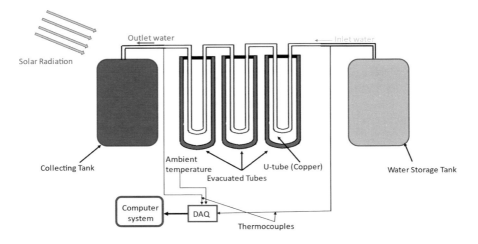

FIGURE 6.4 Schematic diagram of the experimental setup.

during the study: solar radiation intensity, inlet water temperature, outlet temperature, air temperature, atmospheric temperature, and collector inner temperature.

6.3 THERMAL PERFORMANCE EVALUATION

In order to evaluate the thermal performance of a U-tube-type evacuated solar collector, incident solar radiation is measured. The heat produced due to solar radiation is transferred to the working fluid. The actual amount of heat transferred to the fluid is calculated from the expression given below.

$$Q_{net} = \dot{m}_w c_p \Delta T_w \qquad (6.1)$$

where $\dot{m}_w$ is the mass flow rate of the water circulated through the collector, c_p is the specific heat of water, and ΔT_w is the temperature difference between inlet and outlet water of the experimental test section.

6.3.1 Incident Solar Radiation

The amount of solar radiation that falls on the collector is given by the following equation.

$$\text{Incident solar radiation} = A_c I \qquad (6.2)$$

where A_c is the collector surface area and I is the solar radiation intensity.

6.3.2 Daily Thermal Efficiency

Efficiency is the ratio of useful heat gain to the incident solar radiation. Daily thermal efficiency is calculated using the following relation:

$$\text{Efficiency}, \eta = \frac{\text{Useful heat gain}}{\text{Incident solar radiation}} = \frac{m_w c_p \Delta T_w}{A_c I} \qquad (6.3)$$

6.3.3 Uncertainty Analysis

The major sources of errors in various instruments are primarily influenced by systematic errors and random errors. Hence, the uncertainty analysis is performed to predict the measured uncertainties. Maximum uncertainty of ±8% is associated with the calculation of efficiency [13].

Parameter	Devices	Total uncertainty (± %)
Temperature	Thermocouple, K-type	2
Flow rate	Flowmeter	3
Solar radiation	Pyranometer	5
Heat transferred (Q_{net})	-	3.6
Efficiency	-	8

6.4 RESULTS AND DISCUSSION

It is worth mentioning that the experimental work has been performed at NIT Uttarakhand, India, in April 2021. Solar radiation intensity and temperature at various positions have been taken every 15 minutes (after achieving the steady-state condition). Figure 6.5 shows the variation of solar radiation intensity and ambient temperature with respect to time. The maximum recorded solar radiation is 896 W/m^2 tentatively at 12:30 PM. It can be seen that solar radiation is having fluctuating characteristics due to changing climatic and atmospheric conditions. It is obvious that with increasing solar radiation ambient temperature is also increasing. Prevailing fluctuations in the solar intensity are also apparent in the data of ambient temperature. Unlike solar radiation, a maximum ambient temperature of 37.83°C is reported at 2:00 PM. However, it continuously increases with time between 10:00 AM and 2:00 PM.

6.4.1 Temperature Distribution Inside the Collector

In order to measure the temperature inside the evacuated collector, three thermocouples have been placed at an equal distance which provides the average temperature inside the evacuated tubes. Hence, a graph has been drawn between the solar radiation, collector inner temperature, and time. Figure 6.6 clearly illustrates that the maximum temperature is recorded around ≈ 72.5°C. The maximum temperature inside the evacuated tube will be at the core position and the minimum temperature in the entrance section. It is because heat loss by the air will be minimum at the core position compared to the entrance section. Collector temperature is responsible for the heat transfer between the air and the heat transfer fluid (water) circulated in the copper U-tube. Hence, the higher the collector temperature, the higher the outlet

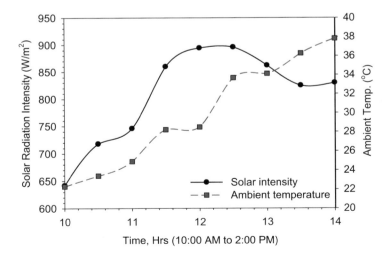

FIGURE 6.5 Variation of solar radiation intensity and ambient temperature w.r.t. time on 18 April 2021.

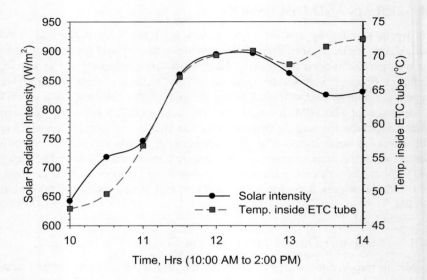

FIGURE 6.6 Variation of collector inner temperature and solar intensity with time.

temperature of the heat transfer fluid. Since the possibility of heat loss is more at the entrance region, therefore, proper insulation is required at the entrance of the evacuated tube so that collector inner temperature would be high and heat loss would be minimum at the entrance point. Figure 6.6 also reveals that the maximum temperature in the ETC is recorded at 2:00 PM. It is in line with the maximum ambient temperature recorded at the same time as discussed in the previous section.

6.4.2 Influence of Coolant Mass Flow Rate

The range of the mass flow rate is being varied from 300 to 1,100 mL/min. Figure 6.7 shows the effect of varying mass flow rate of coolant on the change in temperature of the heat transfer fluid with time.

It should be noted that the temperature difference of the coolant at inlet and outlet has been recorded at varying times. It is clearly depicted that the maximum change in temperature of the water is being recorded as 8.46°C at a mass flow rate of 700 mL/min and as the mass flow rate is being continuously increased, consequently change in temperature of the water is reducing. This is because an increased mass flow rate does not provide sufficient time for the water to capture the collector heat as the velocity of the fluid has been increased. Figure 6.8 illustrates the variation of solar radiation intensity and inlet–outlet temperature difference of the water with respect to mass flow rate. Similar to the earlier discussion, this figure predicts that the maximum recorded outlet temperature difference is 8.46°C at a mass flow rate of 700 mL/min, and corresponding solar radiation is observed around 895 W/m². With an increase in mass flow rate, the outlet temperature of the water is decreasing subsequently, and temperature difference also reduces.

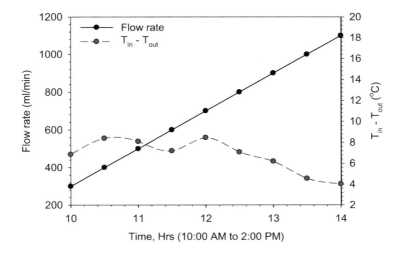

FIGURE 6.7 Influence of mass flow rate on temperature difference with time.

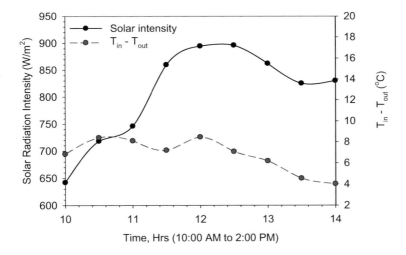

FIGURE 6.8 Effect of solar radiation intensity on the temperature difference of the fluid.

6.4.3 THERMAL PERFORMANCE ASSESSMENT

Thermal efficiency is one of the most appropriate parameters which describe the thermal performance of any thermodynamic system. It has already been mentioned that the efficiency of the ETC is the function of both solar radiation and the mass flow rate. Hence, a graph has been plotted which shows the relation among parameters such as solar radiation intensity, mass flow rate ambient temperature, inlet and outlet temperature of the water, and efficiency. Calculation of efficiency has been done as per Eq. (6.3). Figure 6.9 shows the variation of efficiency and solar radiation intensity with mass flow rate. It is being observed that maximum efficiency is recorded as

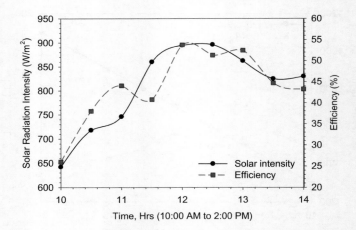

FIGURE 6.9 Variation of solar radiation intensity and efficiency with mass flow rate.

53.76% at a mass flow rate of 700 mL/min and solar radiation intensity of 895 W/m². Furthermore, it is important to note that the efficiency curve shows an almost similar trend of solar radiation. Its maximum value is reported approximately at 12:30 PM where intensity is also high.

6.5 CONCLUSION

A parametric experimental work has been carried out to investigate the heat transfer performance and thermal efficiency of a U-tube-based evacuated tube solar water heater. Variation in the temperature of the cooling medium is mainly recorded with varying solar radiation intensity and mass flow rate. All the experimental runs have been done between 10:00 AM and 2:00 PM at NIT Uttarakhand Srinagar (Garhwal). Experimental results reveal that in the considered range of mass flow rate varied from 300 to 1,100 mL/min, the mass flow rate of 700 mL/min is the most appropriate that favours higher temperature rise of the cooling medium. The maximum outlet temperature of the water is being recorded to 29.97°C at the mass flow rate of 700 mL/min. For the present experimental setup and operating conditions, maximum efficiency of 53.76% is achieved at the solar radiation of 895 W/m².

NOMENCLATURE

A_c	Collector surface area	m²
c_p	Specific heat of water	kJ/kg-K
I	Solar radiation intensity	W/m²
m_w	Mass of water	Kg/s
ΔT_w	Temperature difference of inlet and outlet of the water	°C
η	Efficiency	-

REFERENCES

1. IEA, *World Energy Outlook 2013*. U. S. Energy Information Administration, International Energy Outlook 2013, pp. 01–312, DOE/EIA-0484 (2013), www.eia.gov/ieo/.
2. World Energy Council, "Steering committee," *Cold Reg. Sci. Technol.*, vol. 2, pp. vii–xiv, 1980, Doi: 10.1016/0165-232X(80)90063-4.
3. J. Ghaderian et al., "Performance of copper oxide/distilled water nanofluid in evacuated tube solar collector (ETSC) water heater with internal coil under thermosyphon system circulations," *Appl. Therm. Eng.*, 2017, Doi: 10.1016/j.applthermaleng.2017.04.117.
4. R. Liang, L. Ma, J. Zhang, and L. Zhao, "Performance analysis of a new-design filled-type solar collector with double," *Energy Build.*, vol. 57, pp. 220–226, 2013, Doi: 10.1016/j.enbuild.2012.11.004.
5. W. Wu et al., "Experimental study on the performance of a novel solar water heating system with and without PCM," vol. 171, no. July, pp. 604–612, 2018.
6. X.R. Zhang and H. Yamaguchi, "An experimental study on evacuated tube solar collector using supercritical CO_2," *Appl. Therm. Eng.*, vol. 28, no. 10, pp. 1225–1233, 2008, Doi: 10.1016/j.applthermaleng.2007.07.013.
7. L. Ma, Z. Lu, J. Zhang, and R. Liang, "Thermal performance analysis of the glass evacuated tube solar collector with U-tube," *Build. Environ.*, vol. 45, no. 9, pp. 1959–1967, 2010, Doi: 10.1016/j.buildenv.2010.01.015.
8. R. Liang, L. Ma, J. Zhang, and D. Zhao, "Theoretical and experimental investigation of the filled-type evacuated tube solar collector with U tube," *Sol. Energy*, vol. 85, no. 9, pp. 1735–1744, 2011, Doi: 10.1016/j.solener.2011.04.012.
9. R. Liang, J. Zhang, L. Zhao, and L. Ma, "Research on the universal model of filled-type evacuated tube with U-tube in uniform boundary condition," *Appl. Therm. Eng.*, vol. 63, no. 1, pp. 362–369, 2014, Doi: 10.1016/j.applthermaleng.2013.11.020.
10. Y. Gao et al., "Thermal performance and parameter analysis of a U-pipe evacuated solar tube collector," *Sol. Energy*, vol. 107, pp. 714–727, 2014, Doi: 10.1016/j.solener.2014.05.023.
11. A.K. Pandey, V.V Tyagi, N.A. Rahim, S.C. Kaushik, and S.K. Tyagi, "Thermal performance evaluation of direct flow solar water heating system using exergetic approach," 2015, Doi: 10.1007/s10973-015-4566-4.
12. B.K. Naik, A. Varshney, P. Muthukumar, and C. Somayaji, "Modelling and performance analysis of U type evacuated tube solar collector using different working fluids," *Energy Procedia*, vol. 90, no. December 2015, pp. 227–237, 2016, Doi: 10.1016/j.egypro.2016.11.189.
13. X. Nie, L. Zhao, S. Deng, and X. Lin, "Experimental study on thermal performance of U-type evacuated glass tubular solar collector with low inlet temperature," *Sol. Energy*, vol. 150, pp. 192–201, 2017, Doi: 10.1016/j.solener.2017.04.030.
14. D. Korres and C. Tzivanidis, "A new mini-CPC with a U-type evacuated tube under thermal and optical investigation," *Renew. Energy*, vol. 128, pp. 529–540, 2018, Doi: 10.1016/j.renene.2017.06.054.
15. B.K. Naik, M. Bhowmik, and P. Muthukumar, "Experimental investigation and numerical modelling on the performance assessments of evacuated U – Tube solar collector systems," *Renew. Energy*, vol. 134, pp. 1344–1361, 2019, Doi: 10.1016/j.renene.2018.09.066.
16. D.N. Korres, C. Tzivanidis, I.P. Koronaki, and M.T. Nitsas, "Experimental, numerical and analytical investigation of a U-type evacuated tube collectors' array," *Renew. Energy*, vol. 135, pp. 218–231, 2019, Doi: 10.1016/j.renene.2018.12.003.
17. M. Bhowmik, P. Muthukumar, and R. Anandalakshmi, "Experimental based multilayer perceptron approach for prediction of evacuated solar collector performance in humid subtropical regions," *Renew. Energy*, vol. 143, pp. 1566–1580, 2019, Doi: 10.1016/j.renene.2019.05.093.

18. A.W. Badar, R. Buchholz, and F. Ziegler, "Experimental and theoretical evaluation of the overall heat loss coefficient of vacuum tubes of a solar collector," *Sol. Energy*, vol. 85, no. 7, pp. 1447–1456, 2011, Doi: 10.1016/j.solener.2011.04.001.
19. H. Kim, J. Ham, C. Park, and H. Cho, "Theoretical investigation of the efficiency of a U-tube solar collector using various nano fluids," vol. 94, pp. 497–507, 2016, Doi: 10.1016/j.energy.2015.11.021.
20. Y. Tong, J. Kim, and H. Cho, "Effects of thermal performance of enclosed-type evacuated U-tube solar collector with multi-walled carbon nanotube/water nano fluid," *Renew. Energy*, vol. 83, pp. 463–473, 2015, Doi: 10.1016/j.renene.2015.04.042.
21. B. Xie, C. Li, B. Zhang, L. Yang, G. Xiao, and J. Chen, "Evaluation of stearic acid / coconut shell charcoal composite phase change thermal energy storage materials for tankless solar water heater," vol. 1, no. August 2019, pp. 187–198, 2020, Doi: 10.1016/j.enbenv.2019.08.003.
22. H. Olfian, S. Soheil, M. Ajarostaghi, and M. Farhadi, "Melting and solidification processes of phase change material in evacuated tube solar collector with U-shaped spirally corrugated tube," *Appl. Therm. Eng.*, p. 116149, 2020, Doi: 10.1016/j.applthermaleng.2020.116149.
23. H. Kaya, K. Arslan, and N. Eltugral, "Experimental investigation of thermal performance of an evacuated U-Tube solar collector with ZnO/Etylene glycol-pure water nano fluids," *Renew. Energy*, vol. 122, pp. 329–338, 2018, Doi: 10.1016/j.renene.2018.01.115.
24. M.H. Abokersh, M. El-Morsi, O. Sharaf, and W. Abdelrahman, "On-demand operation of a compact solar water heater based on U-pipe evacuated tube solar collector combined with phase change material," *Sol. Energy*, vol. 155, pp. 1130–1147, 2017, Doi: 10.1016/j.solener.2017.07.008.

Part II

Green Energy Storage

7 Green Technology Solutions for Energy Storage Devices

Himanshu Priyadarshi, Kulwant Singh, and Ashish Shrivastava

CONTENTS

7.1 Introduction 117
7.2 What Is Green Technology? 118
7.3 Role of Green Technology as a Precursor for Energy Storage-Graphene Procurement 120
7.4 Properties of Graphene in the Context of Energy Storage Devices 121
    7.4.1 Graphene's Morphology 121
    7.4.2 Electronic Properties 122
    7.4.3 Optical Properties 123
    7.4.4 Thermal Properties 123
    7.4.5 Mechanical Properties 123
7.5 Role of Machine Learning-Based Prognoses for Green Technology Solutions 124
7.6 Conclusion 125
Acknowledgments 128
References 128

7.1 INTRODUCTION

Far-sighted energy planning is very important for any society desiring to progress peacefully. The sovereignty of a civilization is dependent on whether it is self-sufficient in terms of its energy needs. Adequate reserves of sustainable energy are crucial if a country wants to achieve the millennium development goals. Unfortunately, the wheels of material civilization for many countries are directly dependent on the fossil fuel. This abject dependence has caused huge loss of lives and degradation of life quality metrics. Realization regarding the physiological, economical, ecological, and environmental hazards was an eye-opener for the intelligentsia, for fossil fuel technology has acted as Frankenstein's monster. Green solutions for energy storage will exhibit pertinent advantages as follows: (i) energy management at demand side; (ii) arbitrage of energy assets; (iii) energy assurance; (iv) effective utilization of the

DOI: 10.1201/9781003258209-9

existing infrastructure of the power grid; and (v) cost-cutting by abstention from further investment in creating new power grid infrastructure [1–3].

The pattern of load utilization at the consumer end varies depending upon the total connected capacity at the consumer's end. The policies made by electricity distributing authorities encourage power consumption in a time-segregated manner. Certain time intervals are earmarked as peak hours due to high power consumption; and it is encouraged by the utilities that power consumption may be restricted to vital applications only. It is advisable that non-vital power consumption may be snubbed during peak time. During off-peak hours, when the vital applications are inactive, the desirable and off-peak domestic applications may be operated in order to reduce the burden. The effective implementation of such demand-side management can be done only when the energy can be buffered by dint of energy storage management systems.

The importance of self-sustainability, reliability, and security in terms of energy assets can hardly be overemphasized for the economic development of a country. The pricing of energy in the market varies a lot depending on sources as well as intermediaries in energy harvesting and end-user harnessing. From the entropic vantage point, the most suitable way of energy exchange is in the form of electricity, as the entropy is very low in the electrical form of energy, and hence the efficiency of exchange is high. Although the conventional power grid does not have much scope of making good use of differences in energy pricing, modern smart grids empowered by data-driven intelligence are capable of energy pricing arbitrage i.e., making the best decisions based on differences in energy prices. Empowered by data-driven intelligence, the energy management of the energy storage system can make smart decisions to reap maximum dividends.

The balancing of demand-supply transients was conventionally done by large-inertia alternators. With these high-inertia alternators meeting obsolesce, the onus on dealing with such transients has been shifted to energy storage technology. The time-domain segregation of utilization of electrical power and supply enables the consistency of power supply in the wake of grid blackouts. Thus we have logically established that the success of decentralized renewable energy resources in delivering the reliability of power supply depends on energy storage systems.

7.2 WHAT IS GREEN TECHNOLOGY?

Sciences and technologies are meant to improve the quality of human life on the planet. However, hedonistic psychology has led to the abuse of sciences and technologies, which has not benefited anyone. On the contrary, the natural reaction of harmful technologies has resulted in many man-made disasters, which still haunt many. A case in point is the use of petroleum-powered IC engine vehicles. Although most of us are habituated to using them due to the existing supply chain and logistics, which make this technology so frequently used, the harmful effects of petroleum-powered IC engine vehicles can hardly be overstated.

Starting from the way petroleum is procured by drilling the surface of the earth to the polluting exhausts emitted by the vehicle; the hazards of the petroleum technology can be enumerated as follows: (i) petroleum drilling has been reported to cause earthquakes; (ii) petroleum refineries cause a lot of environmental waste which is quite tough to manage; (iii) the economies of countries which are heavily dependent on petrol and diesel are acutely affected by the fluctuation of the price of petroleum in the international oil market; (iv) oil-rich countries in the Middle East have always been afflicted by war; and (v) the greenhouse gases emitted from the petroleum-powered vehicles denigrates the air quality, making the air unbreathable at many places with a high population density of petroleum-powered vehicles.

One can arguably wonder why did our predecessors fail to see the harmful effects of petroleum-powered vehicles before giving a red-carpet welcome to the petroleum technology for transportation. The petroleum technology has been facilitated to such an extent and integrated with our existence to such an extent that today we are directly dependent on the swings of petroleum price even for our basic amenities. At the time of this chapter being written the price of petroleum is 108 Indian rupees at the retail refilling station, and this has caused great anxieties in the heart of the common populace, robbing their peace of mind.

Whenever a technology has so many harmful effects for everyone, certainly it is not congenial for human existence, and it cannot be said to be a green technology. When we use the term green for a technology, it is very intuitive to understand its meaning. Correlating it with a traffic signal, a vehicle at crossroads before making the decision to move ahead or not looks at the color of the traffic signal. Red color indicates that the vehicle cannot move ahead as it will create collisions with other entities on the road; yellow color indicates the vehicle to be watchful; whereas green color indicates that the vehicle can proceed along the path. Similarly, a technology being referred to as green means that it can propel the vehicle of human societal progress in a smooth manner without accidental hazards.

Now let us understand the context of accidental hazards when a technology is not green enough to be serviceable. The context of hazards may be understood by invoking the quadruple bottom line of corporate social responsibility.

The need to invoke the quadruple bottom line of corporate social responsibility is obvious for a technology that permeates the masses on large scale only by the influence of titanic corporate houses. These corporate organizations are economic overlords influencing the policies of all the policymaker legislations of the world. After the industrial revolution (Industry 1.0), most of the corporate houses were mostly considering short-term profitability as the only reward parameter while evaluating the reward function of any technology, and due to this myopic view, the entire planet has suffered great loss.

Contemporarily, there has been an awakening and realization by dint of the positive examples created by benevolent corporate houses spearheaded by astute and pragmatic leadership that sustainable profitability cannot be obtained at the cost of the following: (i) the planet, which provides all the resources for the sustenance of everyone; (ii) the people, the service providers as well as the client/consumers of the services/product being sold by the business entities; and (iii) the purpose for which the business was conceived. Along with profitability, people, planet, and purpose are considered the four Ps of the quadruple bottom line for corporate social responsibility.

7.3 ROLE OF GREEN TECHNOLOGY AS A PRECURSOR FOR ENERGY STORAGE-GRAPHENE PROCUREMENT

There is tremendous interest among researchers in the exploration of the possibilities for synthesizing graphene-based electrodes for unit cell applications which can be facilely introduced into the industry by the valorization of waste material being dumped in the ecosystem. Inefficient waste handling slashes down the profitability to alarming levels and invites the wrath of environmental protection authorities at multiple aspects such as financial punishment and disrepute. Some examples of energy procurement routes originate from upgradation of lignin (a by-product of cellulose refinery, paper & pulp industry), coal, agrarian by-products, waste fungal substrate, waste coffee, etc.

A careful observation of the starting materials that produce most of the industrial waste (or by-products) based graphene electrodes we will be discussing herein, are organic polymers in essence. For instance, polysaccharides perform structural as well as storage functionality in the material bodies of the living entities. Mahmood et al. [4] strongly advocated for laser-scribing as a cost-efficient, controllable, scalable, efficient technique for the procurement of unit cell electrolyte grade graphene from the lignin which is and will be abundantly available as a by-product from the prospective cellulose bio-refinery as well paper & pulp industry.

Also, Chu et al. [5] exhibited a lot of belief in economically lean and ecologically green high-volume production of highly porous graphene from crude polysaccharide (procurable from fungal waste) with commendable specific capacitance using a three-electrode system, in an attempt to answer the impending energy challenges aggravated by the issues of fast energy retrieval by the energy sinks from the energy sources [6,7].

It has also been reported [8] that porous carbon, derived from a really simple treatment of coconut shell, can be used as an electrode material without any decoration for the sake of conductivity enhancement for ultracapacitor applications by dint of exercising optimization controllability over mesopore to micropore distribution through variation in the activation parameters such as temperature and water flow rate, resulting in desired conductivity, higher energy density & durable cyclic stability even in the activated porous carbon directly prepared from the coconut shell.

The raw materials pertaining to the fabrication of super-capacitors play a very important role in end-user cost affordability [9–14]. Most of the time the procurement of the precursors for the graphene is a difficult task in itself. The need of creating a problem-solving platform is being acutely felt in this context, with the advent of Industry 4.0; the utilization of industrial waste can be very easy, as this fourth-generation industrial revolution is touted to create an integrated problem-solving platform for the waste to wealth conversion.

This is the motto behind presenting a consolidated review of the contributions made by the researchers toward converting the so-called waste from different industries into the wealth of energy. Such an approach seems to be very promising when evaluated from the multiple-perspectives such as environmental protection, life-cycle costing or cradle-to-grave analysis of the materials, cost-benefit analysis for industries, and will surely help in converting loss-making ventures by tangibly increasing their profitability reflected in their balance sheets, with very little investment in waste-management agencies [15]. This will also help such power sector industries [16] in carving/upgrading their brand equity by boosting their compliance with the quadruple bottom line of corporate social responsibility.

A careful observation reveals that the precursors used for arriving at the graphene electrode for ultracapacitor applications is through the route of porous activated carbon. Carbon happens to be the elementary constituent for agricultural wastes such as rice husk, black gram husk, and green gram husk [17–19]. These husks constitute one-third of the corresponding agricultural produce and are left unattended after crop cultivation of these crops which are abundantly cultivated in Asia, except for their sporadic unorganized utility as cattle-feed, bio-fertilizer, and domestic fuel. The sheer magnitude of this agro-waste hints at tapping their unutilized potential of being a source of activated carbon [20–33] which can be further processed by dint of adsorption to be used as a graphitic electrode for unit cell applications. Once the cyclic stability and the capacitance retentivity of these unit cell electrodes reach their expiry date, these can be recycled for various other applications like wastewater treatment with nil sludge formation by adsorption of phenol on reactivated carbon. Thus, what we are attempting is to contribute toward a complete ecosystem with avenues for an industrial partnership where all partners can collaborate to thrive in terms of profitability.

7.4 PROPERTIES OF GRAPHENE IN THE CONTEXT OF ENERGY STORAGE DEVICES

Graphene is hailed as the two-dimensional magic material for energy storage devices intended for swift power transactions [34,35]. The evidence that follows in this section will present facts that will advocate for intensifying the focus on graphene research. The various attempts at synthesis, characterization, and investigation of the structural, morphological, electronic, thermal, optical, mechanical, and application-affiliated properties of graphene have been very encouraging toward utilizing it for energy harnessing. Although these results are still at the laboratory reportage level, a skeptic approach inhibits the entrepreneurial approach to the tagging of graphene for unit cell design due to the limitations on repeatability and scalability. Despite the differences in the properties reported by myriad research groups, still it is important in our context to review the properties from an optimistic perspective, one by one.

7.4.1 Graphene's Morphology

The morphology of graphene after being discovered has decimated the popular belief system among researchers in material science that two-dimensional atomic crystals do not exist only for any finitely quantified temperatures [34,35]. The convincing empirical evidence mustered, reported, and systematically elucidated [36–38], which establishes that graphene exists as a long-range ordering of two-dimensional atomic crystals. There are many misconceptions regarding graphene still lingering in the academic circles; many times it is enquired as to how it can be justified to be a magic material, which is solely by dint of its properties, and these properties stem from its morphology. Graphene is not an entity which can be used independently, for the long-range ordering of this two-dimensional structure is unusable because of being unstable as a free entity. Graphene is a single atom thick sheet of sp2-hybridized carbon atoms which utilize carbon–carbon double bonds, and it may be considered as the elementary building block of carbon materials and is the most fundamental analytic derivative obtained from graphene. The morphological definition of graphene may be presented as follows: a monoatomic bi-dimensional sheet of sp2-hybridized orbital carbon atoms bonded together in a hexagonal fashion. The stability of graphene has been attributed to its curvilinear surface, which has been confirmed by the proportionate increase in the breadth of diffraction peaks (in the X-ray diffraction fingerprint of the graphene) with the increase in tilt angle; this is in stark contrast with the crystalline behavior observed in three-dimensional crystal. Another important structural asset other than the surface is the edge state of the graphene, which has been found to correlate with its microstructural properties. Research groups have focused on increasing the edge states. Different variants of graphene morphologies repurposed for energy storage application from a multitude of structural precursors like carbon microtubings, microfibers, carbon nanotubes, and so on, wherein various morphologies have been studied [39–46].

7.4.2 Electronic Properties

The premium quality two-dimensional atomic crystalline morphology of graphene bestows it with superlative electronic properties. The electron mobility reported for graphene is so high that it can be considered as good as metallic, and hence graphene finds application in current collectors in energy storage devices. Graphene sheets have been experimentally proven to have high carrier mobility for carriers of either polarity. From an energy band-gap perspective, graphene is said to be a zero-energy band-gap material. In fact, there is a singular point intersection between conduction band and valence band, which is called the Dirac point. This Dirac point is the reason corroborated for the zero-energy band gap in graphene. The density of states for graphene is null at the Dirac point, and first-order dispersion relations conform to zero-density-of-states effective mass. Therefore, the electrons in graphene's two-dimensional atomic crystal manifest as massless Dirac fermions, which causes ultra-high electron mobility. The carrier transport is ultra-ballistic on the micrometric scale by dint of the superlative carrier mobility reported in the range of 15,000–200,000 cm^2/Vs, with remarkably weak temperature dependence. The lower limit of mobility has been reported with very high carrier congestion up to 10^{13} cm^{-3}, whereas the upper extremum has been reported for mechanically exfoliated, substrate-uninfluenced, scaffolded monolayer graphene [47–50]. Graphene is also endowed with amphoteric electric field effect by dint of which the carriers can be flexibly modulated between electrons as well as holes. This tenability of the carrier transport in graphene can be modulated by dint of its amenability with substrates for specific applications.

7.4.3 OPTICAL PROPERTIES

The peculiarly workable properties from an optoelectronics and photovoltaic energy storage vantage point for the single-layer graphene corroborates with its abnormally negligible energy states borne out of its chirality. It has got very high opacity and graphene sheets of up to ten atomic layers have been demonstrated to absorb 2.3% of the white light [51–55]. However, this utilizable opacity is manifest only up to ten atomic layers. Hence, graphene film can be advantageously engineered for photovoltaic energy conversion as well as photo-detection. In our context of electric vehicle applications, it is noteworthy that graphene film solar cells can be utilized for harnessing solar energy and replenishing graphene-based energy storage devices [50,56–58].

7.4.4 THERMAL PROPERTIES

Terminal hazards resulting due to thermal runaway avalanching due to sporadic hotspots have been an infamous issue in energy storage devices, nay all electrical devices. Niggardly levels of thermal conductivity cause many internal device malfunctioning – filament shorting within the cell assembly of the lithium-ion batteries just being a case in point. The insulation requirements of the EV can also be catered to some extent by using graphene-based nanocomposites and thermal pastes, apart from the thermal management of state-of-the-art electronic chips. Thermal design considerations are important during delivery as well as uptake of energy from the storage devices, for an efficient design will not have much tolerance for heat losses. Apart from heat loss minimization, it is also mandatory that heat loss has to be innocuously channelized without transforming into any device hazards; hence the ultra-high thermal conductivity will be vital for secure thermal management.

Monolayer graphene has been reported to be the highest intrinsic thermal conductivity ever reported [59,60] up to 6,000 W/mK, which is a very important attribute in the design of energy storage devices. However, the aforementioned thermal conductivity has not taken into account the influence of substrate, which is generally the case for applications; on the contrary, the substrate-scaffolded graphene's thermal conductivity has been reported to be as low as 600 W/mK. It is noteworthy that electronic contribution to thermal conductivity is negligible due to the comparatively lower carrier density of virgin graphene [61–63]. Connectedly, it has been inferred that the thermal conductivity is facilitated by ballistic transport at low temperature and gradient-based diffusion of phonons at sufficiently high temperatures. The substrate-specific degradation of the thermal conductivity is due to interfacial phonon leakage and scattering of vibrational modes.

7.4.5 MECHANICAL PROPERTIES

The high strength of the bond between carbon atoms imparts stupendous strength to the graphene structure while allowing it to remain flexible. The humongous elasticity as validated from the stress–strain experiments conducted by various research groups has opened a plethora of applications in the domain of flexible electronics, bendable gadgetry, wearable energy storage devices, and so on. The seamless integration of electronics with the apparel can be ably supported by the dint of mechanical properties of graphene. The pristine graphene has been experimentally acknowledged to bear immensely high elastic strength (Young's modulus of 1 terapascal) as well as very high intrinsic fracture strength (130 gigapascal) by dint of nano-indentation technique using atomic force microscopy. However, graphene paper manufactured from graphene oxide has been reported to have Young's modulus which is 3.2% of that of pristine graphene, while the yield strength gets reduced to 0.1%. Chemically reduced graphene procured from reduced graphene oxide in hydrogen plasma ambience has yielded results closer to the virgin graphene and achieves up to 25% of the average Young's modulus [64–67]. These can be attributed to the defects caused due to various processes.

The aforementioned flexibility along with mechanical strength can be very easily utilized for energy harvesting by dint of transducers, apart from lending itself toward the manufacturing of deformable unit cells as well as batteries.

7.5 ROLE OF MACHINE LEARNING-BASED PROGNOSES FOR GREEN TECHNOLOGY SOLUTIONS

The asset-light model has evolved as one of the most important factors for successful start-up ventures in the field of the energy sector wherein the smart entrepreneurs do not want to get suffocated with heavy infrastructural investments. The focus of energy sector start-up ventures has shifted toward creating a profitable problem-solving platform with the least investment in infrastructure. Intelligent energy storage systems have been pivotal to avoid the deployment of infrastructure unless it is vital. This has been experienced not only in the generation, transmission, and distribution but also holds true at the energy utilization premises of the consumer. Renewable energy resources have been the driving impetus for the decentralization of the otherwise centralized power grid. However, renewable energy resources are heavily dependent on the vagaries of the weather, and hence it is important that distributed generation is augmented with energy storage systems which can hold the energy when surplus, and release it when there is a deficit.

Solar energy is a vital source for the sustenance of life on the planet earth. Life is simply unimaginable without solar energy. Hence it is impossible to overemphasize the importance of technological endeavors for maximizing the utilization of the energy being bestowed upon us by dint of solar radiation. With the efforts of increasing the penetration of distributed generation over centralized generation, the research in solar energy has got even more impetualized. The need of backing up the efforts of solar photovoltaic enabled decentralization of the power grid by intelligent energy storage systems can hardly be overstated. The processing unit of the energy storage systems must be endowed with a dynamic as well historical cognizance of the energy being fed to it by the solar photovoltaic panels for different time intervals. Dynamic cognizance is of course dependent on the sensors associated with the system. The term historical cognizance has been used in the context that the intelligence of the energy storage management system machine is trained with such a huge dataset, that it is aware of the expected insolation on a yearly, monthly, and daily basis. The historical data of the weather at different localities can be used for bestowing historical cognizance on the processor of the energy storage management system. The decidability of the solar radiation for different time-frames is non-deterministic as per the theory of computation; i.e., there is no computer algorithm which can definitely state the insolation at a given latitude and longitude on a given time. Although the sun is fixed in its trajectory, and ultra-punctual in distributing its radiation on the earthly plane, the climate and the weather at a locality are highly dependent on the nature of civilization at a particular locality, which remains stochastic – these stochastic factors include temperature, humidity, dust vulnerability, pollution, and so on. Again we restate our point that responsible environmental behavior is very important for a civilization in order to harness renewable sources of energy. In order to predict the insolation, one needs to resort to deep learning-based prediction techniques [68–71].

Albeit, scholars may debate whether machine learning is a sub-domain of the broader domain of artificial intelligence, it won't be inappropriate to say that machine learning is the driving technology for constructive dependable models based on historical data. There are many mathematical techniques which provide very effective ways of solving the data-driven prognoses problem, but it will be beyond the scope and intent of this chapter to illustrate all of them here.

The causal prediction procedure for the solar energy received on the basis of the historical data can be implemented by using regression learner tools available in different machine-learning packages. A very simple category of machine-learning techniques is linear regression. The simplicity of linear regression is due to the fact that it is based on linear curve fitting for the available dependents and functions. This curve fitting is done using the Taylor series approximation. The regression exercise starts with assigning random values to the function's parameters, iteratively compares the predicted value with the ground truth, finds the error, and updates the weights and biases using backpropagation. Some elementary variants of linear regression have been represented in the following. The data set used herein has been made publicly available by the Malaviya National Institute of Technology situated at Jaipur. In our demonstration of how machine learning can help predict the insolation, we have kept the algorithm very simple. Figures 7.1 and 7.2 illustrate the prediction with all components intact, and Figures 7.3 and 7.4 illustrates the prediction using the principal component analysis keeping the attributes with only 95% variance.

7.6 CONCLUSION

For more than a century now, the entire planet as well as the inhabitants of the planet has been suffering because the economic overlords failed to adopt a vision that would have ensured sustainable progress in harmony with the people as well as the planet. The reason may be due to lack of purpose, and thus capitalists spent their fiscal capital in vain search of profitability. The purpose of business implementation of a technology has to be properly aligned with people as well as the planet; such an approach will naturally ensure profitability.

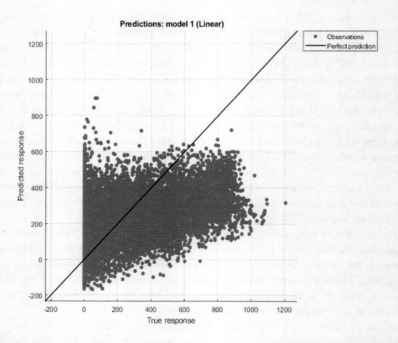

FIGURE 7.1 Predicted response versus true response using a deeply learnt linear regression machine without ignoring any component.

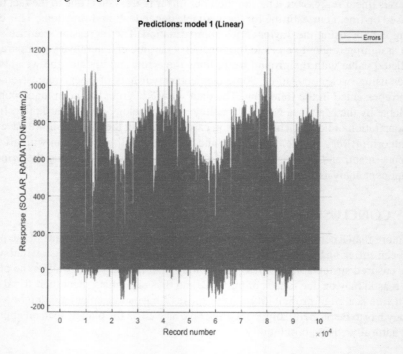

FIGURE 7.2 Variation of response with respect to different weather data tuples using a deeply learnt linear regression machine without ignoring any component.

Green Technology Solutions

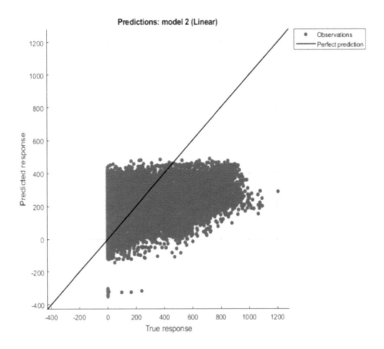

FIGURE 7.3 Predicted response versus true response using a deeply learnt linear regression machine using principal component analysis.

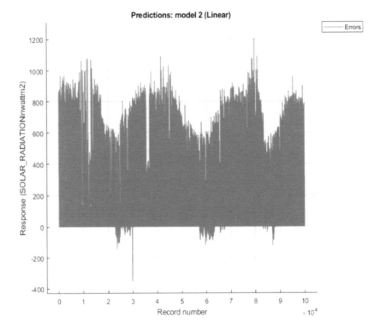

FIGURE 7.4 Variation of response with respect to different weather data tuples using a deeply learnt linear regression machine with principal component analysis.

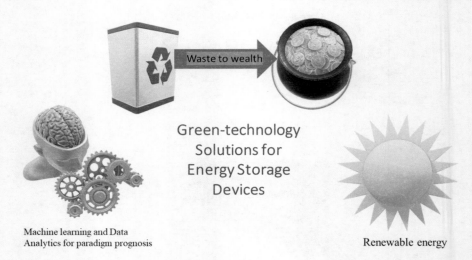

FIGURE 7.5 Graphical representation of the essence of green technology solutions for energy storage devices.

The implementation of sustainable profitability has to be ensured in the planning stage much before the business proliferation of the technology; a suitable tool to this effect is life-cycle costing or cradle-to-grave analysis. The cradle-to-grave analysis is a term which is inspired by Mother Nature in her excellence in the design of different types of creations which exist in perfect harmony with each other. Life-cycle costing of a technology or product is done to ensure whether it is green or not in the context discussed in the foregoing. This technique involves cost-benefit analysis at different stages of the technologies' life cycle, which gives a correlation between the cost involved and the benefits obtained. The cost has to include environmental costs as well.

With this cognizance, technology can be surely utilized for the betterment of the entire world by making energy green and affordable for everyone. The entire approach can be driven by utilizing waste for energy storage materials and using machine learning for better decision-making (Figure 7.5).

ACKNOWLEDGMENTS

We take this opportunity to express our gratitude to everyone who has directly or indirectly contributed to this work. Special gratitude is due to our teachers who have given us the motivation of lifelong learning, and also to our leadership at Manipal University Jaipur, as well as Malaviya National Institute of Technology, Jaipur.

REFERENCES

1. Deng, J., Bae, C., Denlinger, A. and Miller, T. Electric vehicles batteries: Requirements and challenges. *Joule* (2020), Doi: 10.1016/j.joule.2020.01.013.

2. Augustyn, V., McDowell, M.T., and Vojvodic, A. Toward an atomistic understanding of solid-state electrochemical interfaces for energy storage. *Joule* (2018), Doi: 10.1016/j.joule.2018.10.014.
3. Xu, L., Tang, S., Cheng, Y., Wang, K., Liang, J., Liu, C., Cao, Y.-C., Wei, F., and Mai, L. Interfaces in solid-state lithium batteries, *Joule* (2018) 2, 1991–2015.
4. Pumera, M. Graphene-based nanomaterials for energy storage. *Energy Environ. Sci.* (2011) 4, 668–674.
5. The Energy and Resources Institute. Global oil markets and India's vulnerability to oil shocks. TERI-NFA Working Paper Series No. 18 (2014). http://www.teriin.org/projects/nfa/ pdf/working-paper-No18-Oil-volatility.pdf.
6. Lutsey, N., Searle, S., Chambliss, S., and Bandivadekar, A. *Assessment of Leading Electric Vehicle Promotion Activities in United States Cities.* Washington, DC: ICCT (2015). http://www.theicct.org/sites/default/ files/publications/ICCT_EV-promotion-UScities_20150729.pdf.
7. Emission factors estimated based on petroleum product distribution emissions from pipeline, road, and rail modes as reported by Hindustan Petroleum Corporation Limited in Expert Workshop organized by Petroleum Federation of India in October (2015).
8. Xu, C., Xu, B., Gu, Y., Xiong, Z., Sun, J., and Zhao, X.S. Graphene-based electrodes for electrochemical energy storage. *Energy Environ. Sci.* (2013) 6, 1388–1414.
9. Mahmood, F., Zhang, C., Xie, Y., Stalla, D., Lin, J., and Wan, C. Transforming lignin into porous graphene via direct laser writing for solid-state supercapacitors. Doi: 10.1039/c9ra04073k.
10. Chu, M., Li, M., Han, Z., Cao, J., Li, R., and Cheng, Z. Novel biomass-derived smoke-like carbon as a supercapacitor electrode material. Doi: 10.1098/rsos.190132.
11. Conway, B. E. *Electrochemical Supercapacitors: Scientific Fundamentals and Technological Applications.* New York: Plenum Publishers, (1999).
12. Stoller M.D., Park S., Zhu Y., An J., and Ruoff R.S. Graphene based ultracapacitors. *Nanoletters* (2008) 8(10), 3498–3502.
13. Mi, J., Wang, X.R., Fan, R.J., Qu, W.H., and Li, W.C. Coconut-shell-based porous carbons with a tunable micro/mesopore ratio for high-performance supercapacitors. *Energy Fuels* (2012) 26, 5321–5329.
14. Winter, M., and Brodd, R.J. What are batteries, fuel cells, and supercapacitors? *Chem. Rev.* (2004) 104, 4245–4270.
15. Pandolfo, A.G., and Hollenkamp, A.F. Carbon properties and their role in supercapacitors. *J. Power Sourc.* (2006) 157, 11–27.
16. Hingorani, N.G. Introducing custom power. *IEEE Spect.* (1995) 32, 41–48.
17. Srihari, V., and Das, A. Comparative studies on adsorptive removal of phenol by three agro-based carbons: Equilibrium and isotherm studies. *Ecotoxicol. Environ. Safety* (2008) – Elsevier.
18. Simon, P., and Gogotsi, Y. Materials for electrochemical capacitors. *Nature: Mater.* (2008) 7, 845–854.
19. Kotz, R. Principles and applications of electrochemical capacitors. *Electrochimica Acta* (2000) 45, 2483–2498.
20. Miller, J.R., and Burke, A.F. Electrochemical capacitors: Challenges and opportunities for real-world applications. *Electrochem. Soc. Interf.* (2008) 17, 53–57.
21. Gamby, J., Taberna, P.L., Simon, P., Fauvarque, J.F., and Chesneau, M. Studies and characterization of various activated carbons used for carbon/carbon supercapacitors. *J. Power Sourc.* (2001) 101, 109–116.
22. Shi, H. Activated carbons and double layer capacitance. *Electrochim. Acta* (1995) **41**, 1633–1639.
23. Qu, D., and Shi, H. Studies of activated carbons used in double-layer capacitors. *J. Power Sourc.* (1998) 74, 99–107.

24. Qu, D. Studies of the activated carbons used in double-layer supercapacitors. *J. Power Sourc.* (2002) 09, 403–411.
25. Kim, Y.J., Horie, Y., Ozaki, S., Matsuzawa, Y., Suezaki, H., Kim, C., Miyashita, N., and Endo, M. Correlation between the pore and solvated ion size on capacitance uptake of PVDC-based carbons. *Carbon* (2004) 42, 1491.
26. Izutsu, Kosuke. *Electrochemistry in nonaqueous solutions*. John Wiley & Sons, 2009.
27. Fernandez, J.A. et al., Performance of mesoporous carbons derived from poly(vinyl alcohol) in electrochemical capacitors. *J. Power Sourc.* (2008) 175, 675.
28. Jurewicz, K., Morishita, T., Toyoda, M., Inagaki, M., Stoeckli, F., and Centeno, T.A. Capacitance properties of ordered porous carbon materials prepared by templating procedure. *J. Phys. Chem. Solids* (2004) 65, 287.
29. Fuertes, A.B., Lota, G., Centeno, T.A., and Frackowiak, E. Templated mesoporous carbons for supercapacitor application. *Electrochim. Acta* (2005) 50, 2799.
30. Salitra, G., Soffer, A., Eliad, L., Cohen, Y., and Aurbach, D. Carbon electrodes for double-layer capacitors. I. Relations between ion and pore dimensions. *J. Electrochem. Soc.* (2000) 147, 2486–2493.
31. Vix-Guterl, C., Frackowiak, E., Jurewicz, K., Friebe, M., Parmentier, J., and Béguin, F. Electrochemical energy storage in ordered porous carbon materials. *Carbon* (2005) 43, 1293–1302.
32. Mermin, N.D. Crystalline order in two dimensions. *Phys. Rev.* (1968) 176, 250.
33. Mermin, N.D., and Wagner, H. Absence of ferromagnetism or antiferromagnetism in one- or two-dimensional isotropic Heisenberg models. *Phys. Rev. Lett.* (1966) 17, 1133–1136.
34. Novoselov, K.S., Geim, A.K., Morozov, S.V., Jiang, D.E., Zhang, Y., Dubonos, S.V., Grigorieva, I.V., and Firsov, A.A. Electric field effect in atomically thin carbon films. *Science* (2004) 306, 666–669.
35. Novoselov, K.S., Jiang, D., Schedin, F., Booth, T.J., Khotkevich, V.V., Morozov, S.V., and Geim, A.K. Two-dimensional atomic crystals. *Proc. Natl. Acad. Sci. U.S.A.* (2005) 102, 10451–10453.
36. The Nobel Prize in Physics. Nobelprize.org. Nobel Media AB 2014. Web. 23 Sep 2014 (2010). Available at http://www.nobelprize.org/nobel_prizes/physics/laureates/2010/.
37. Bianco, A. et al., All in the graphene family—A recommended nomenclature for two-dimensional carbon materials. *Carbon* (2013) 65, 1–6.
38. Meyer, J.C., Geim, A.K., Katsnelson, M.I., Novoselov, K.S., Booth, T.J. and Roth, S. The structure of suspended graphene sheets. *Nature* (2007) 446, 60–63.
39. Peng, J., Gao, W., Gupta, B.K., Liu, Z., Romero-Aburto, R., Ge, L., Song, L., Alemany, L.B., Zhan, X., and Gao, G. Graphene quantum dots derived from carbon fibers. *Nano Lett.* (2012) 12, 844–849.
40. Kosynkin, D.V., Higginbotham, A.L., Sinitskii, A., Lomeda, J.R., Dimiev, A., Price, B.K., and Tour, J.M. Longitudinal unzipping of carbon nanotubes to form graphene nanoribbons. *Nature* (2009) 458, 872.
41. Kim, H.-K., Bak, S.M., Lee, S.W., Kim, M.S., Park, B., Lee, S.C., Choi, Y.J., Jun, S.C., Han, J.T., and Nam, K.W. Scalable fabrication of micron-scale graphene nanomeshes for high-performance supercapacitor applications. *Energy Environ. Sci.* (2016) 9, 1270–1281.
42. Cheng, H., Huang, Y., Shi, G., Jiang, L., and Qu, L. Graphene-based functional architectures: Sheets regulation and macrostructure construction toward actuators and power generators. *Acc. Chem. Res.* (2017) 50, 1663–1671.
43. Meng, Y., Zhao, Y., Hu, C., Cheng, H., Hu, Y., Zhang, Z., and Qu, L. All-graphene core-sheath microfibers for all-solid-state, stretchable fibriform supercapacitors and wearable electronic textiles. *Adv. Mater.* (2013) 25, 2326–2331.

44. Hu, C., Zhao, Y., Cheng, H., Wang, Y., Dong, Z., Jiang, C., and Qu, L. Graphene microtubings: Controlled fabrication and site specific functionalization. *Nano Lett.* (2012) 12, 5879–5884.
45. Zhao, Y., Hu, C., Hu, Y., Cheng, H., Shi, G., and Qu, L. A versatile, ultralight, nitrogen-doped graphene framework. *Angew. Chem. Int. Ed.* (2012) 124, 11533–11537.
46. Ye, M., Dong, Z., Hu, C., Cheng, H., Shao, H., Chen, N., and Qu, L. Uniquely arranged graphene-on-graphene structure as a binder-free anode for high-performance lithium-ion batteries. *Small* (2014) 10, 5035–5041.
47. Yang, L., Park, C.H., Son, Y.W., Cohen, M.L. and Louie, S.G. Quasiparticle energies and band gaps in graphene nanoribbons. *Phys. Rev. Let.* (2007) 99, 186801.
48. Son, Y.-W., Cohen, M.L., and Louie, S.G. Half-metallic graphene nanoribbons. *Nature* (2006) 444, 347–349.
49. Girit, Ç.Ö. et al., Graphene at the edge: Stability and dynamics. *Science* (2009) 323, 1705–1708.
50. Ma, T., Ren, W., Zhang, X., Liu, Z., Gao, Y., Yin, L.C., Ma, X.L., Ding, F., and Cheng, H.M. Edge-controlled growth and kinetics of single crystal graphene domains by chemical vapor deposition. *Proc. Natl. Acad. Sci. U.S.A.* (2013) 110, 20386–20391.
51. Yang, Y., Liu, X., Zhu, Z., Zhong, Y., Bando, Y., Golberg, D., Yao, J., and Wang, X. The role of geometric sites in 2D materials for energy storage. *Joule* (2018), Doi: 10.1016/j.joule.2018.04.027.
52. Song, H., Liu, J., Liu, B., Wu, J., Cheng, H.M., and Kang, F. Two-dimensional materials for thermal management applications. *Joule* (2018), Doi: 10.1016/j.joule.2018.01.006.
53. Bolotin, K.I., Sikes, K.J., Jiang, Z., Klima, M., Fudenberg, G., Hone, J., Kim, P. and Stormer, H.L. Ultrahigh electron mobility in suspended graphene. *Solid State Commun.* (2008) 146, 351–355.
54. Nair, R. R., Blake, P., Grigorenko, A. N., Novoselov, K.S., Booth, T.J., Stauber, T., Peres, N.M., and Geim, A.K. Fine structure constant defines visual transparency of graphene. *Science* (2008) 6, 1308.
55. Kuzmenko, A. B., Van Heumen, E., Carbone, F., and Van Der Marel, D. Universal optical conductance of graphite. *Phys. Rev. Lett.* (2008) 100, 117401.
56. Malard, L.M., Pimenta, M.A., Dresselhaus, G., and Dresselhaus, M.S. Raman spectroscopy in graphene. *Phys. Rep.* (2009) 473, 51–87.
57. Oostinga, J.B., Heersche, H.B., Liu, X., Morpurgo, A.F., and Vandersypen, L.M. Gate-induced insulating state in bilayer graphene devices. *Nat. Mater.* (2008) 7, 151–157.
58. Bae, S. et al., Roll-to-roll production of 30-inch graphene films for transparent electrodes. *Nat. Nanotech.* (2010) 5, 574–578.
59. Balandin, A.A., Ghosh, S., Bao, W., and Calizo, I. Superior thermal conductivity of single layer graphene. *Nano Lett.* (2008) 8, 902–907.
60. Yu, C., Shi, L., Yao, Z., and Majumdar, A. Thermal conductance and thermopower of an individual single-wall carbon nanotube. *Nano Lett.* (2005) 5, 1842–1846.
61. Seol, J. H. et al., Two-dimensional phonon transport in supported graphene. *Science* (2010) 328, 213–216.
62. Pop, E., Mann, D., Wang, Q., Goodson, K. and Dai, H. Thermal conductance of an individual single wall carbon nanotube above room temperature. *Nano Lett.* (2006) 6, 96–100.
63. Berber, S., Kwon, Y.-K. and Tománek, D. Unusually high thermal conductivity of carbon nanotubes. *Phys. Rev. Lett.* (2000) 84, 4613–4616.
64. Lee, C., Wei, X., Kysar, J.W., and Hone, J. Measurement of the elastic properties and intrinsic strength of monolayer graphene. *Science* (2008) 321, 385–388.
65. Chen, H., Müller, M.B., Gilmore, K.J., Wallace, G.G., and Li, D. Mechanically strong, electrically conductive, and biocompatible graphene paper. *Adv. Mater.* (2008) 20, 3557–3561.

66. Van Lier, G., Van Alsenoy, C., Van Doren, V., and Geerlings, P. Ab initio study of the elastic properties of single-walled carbon nanotubes and graphene. *Chem. Phys. Lett.* (2000) 326, 181–185.
67. Dikin, D.A., Stankovich, S., Zimney, E.J., Piner, R.D., Dommett, G.H., Evmenenko, G., Nguyen, S.T. and Ruoff, R.S. Preparation and characterization of graphene oxide paper. *Nature* (2007) 448, 457–460.
68. Patel, R., and Pagalthivarthi, K.V. MATLAB-based modelling of power system components in transient stability analysis. *Int. J. Model. Simulat.* (2005) 25(1), 43–50.
69. Patel, R., and Mittal, A. *Programming in MATLAB - A Problem Solving Approach.* India: Pearson, (2014).
70. Mittal, A., and Kassim, A. *Bayesian Network Technologies: Applications and Graphical Models: Applications and Graphical Models.* IGI global, (2007).
71. Mittal, A., and Cheong, L.H. Addressing the problems of Bayesian network classification of video using high-dimensional features. *IEEE Trans. Knowl. Data Eng.* (2004, March) 16(2), 230–244.

8 Computational Fluid Dynamic Simulation of Thermal Convection in Green Fuel Cells with Finite Volume and Lattice Boltzmann Methods

O. Anwar Bég, Hamza Javaid, Sireetorn Kuharat,
Ali Kadir, Henry J. Leonard, Walid S. Jouri,
Tasveer A. Bég, V.R. Prasad, Z.
Ozturk, and Umar F. Khan

CONTENTS

8.1	Introduction	134
8.2	ANSYS FLUENT CFD Model	141
	8.2.1 Mass Conservation Equation	141
	8.2.2 Momentum Conservation Equation	142
8.3	Convergence Study	147
8.4	Numerical Results, Visualization, and Discussion	147
	8.4.1 Case I: PEM Fuel Cell	160
	8.4.2 Case II: AFC/PAFC Fuel Cell	161
	8.4.3 Case III: SOFC Fuel Cell	163
8.5	Further Validation with Thermal LBM Code	165
8.6	Conclusions	168
References		169

8.1 INTRODUCTION

A fuel cell is a device that converts chemical potential energy (energy stored in molecular bonds) into electrical energy. A PEM (proton-exchange membrane) cell uses hydrogen gas (H_2) and oxygen gas (O_2) as fuel. The products of the reaction in the cell are water, electricity, and heat. This is a substantial improvement over internal combustion engines, coal-burning power plants, and nuclear power plants, all of which produce harmful by-products. Since O_2 is readily available in the atmosphere, it is merely required to supply the fuel cell with H_2 which can come from an electrolysis process [1]. Many advantages are achieved with PEM fuel cells. By converting chemical potential energy directly into electrical energy, fuel cells avoid the "thermal bottleneck" (a consequence of the second law of thermodynamics) and are thus inherently more efficient than combustion engines, which must first convert chemical potential energy into heat, and then mechanical work. Also, direct emissions from a fuel cell vehicle are just water and a little heat. This is a marked improvement over the internal combustion engine's litany of greenhouse gases. Fuel cells have no moving parts. They are thus much more reliable than traditional engines and circumvent the need for continuous maintenance and diagnostics [2]. An example of the mechanism of a PEM fuel cell is shown in Figure 8.1.

There are four fundamental components in a PEM fuel cell [3]: *anode*, *cathode*, *electrolyte*, and *catalyst*. The anode, the negative post of the fuel cell, has several functions. It conducts the electrons that are freed from the hydrogen molecules so that they can be used in an external circuit. It has channels etched into it that disperse the hydrogen gas equally over the surface of the catalyst. The cathode, the positive post of the fuel cell, has channels etched into it that distribute the oxygen to the surface of the catalyst. It also conducts the electrons back from the external circuit to the catalyst, where they can recombine with the hydrogen ions and oxygen to form water. The electrolyte is PEM. This specially treated material only conducts *positively charged ions*. The membrane blocks electrons. For a PEM fuel cell, the membrane must be hydrated in order to function and remain stable. Finally, the catalyst is a special material that facilitates the reaction of oxygen and hydrogen. It is usually made of platinum nanoparticles very thinly coated onto carbon paper or cloth. The catalyst is rough and porous so that the maximum surface area of the platinum can be exposed to the hydrogen or oxygen. The platinum-coated side of the catalyst faces the PEM. As the name implies, the heart of the cell is PEM. It allows protons to pass through it virtually unimpeded, while electrons are blocked. When the H_2 hits the catalyst and splits into protons and electrons (a proton is the same as an H^+ ion) the protons go directly through to the cathode side, while the electrons are forced to travel through an external circuit. Along the way they perform useful work (e.g. lighting a bulb or driving a motor) before combining with the protons and O_2 on the other side to produce water. Pressurized hydrogen gas (H_2) enters the fuel cell on the anode side. This gas is forced through the catalyst by the pressure. When an H_2 molecule comes into contact with the platinum on the catalyst, it splits into two H^+ ions and two electrons (e^-). The electrons are conducted through the anode, where they make their way through the external circuit (doing useful work

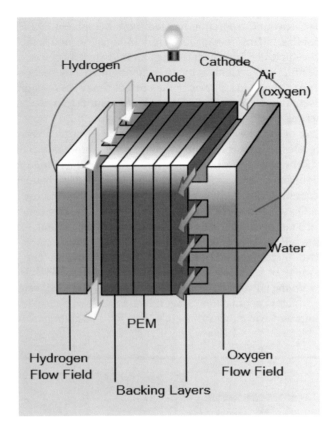

FIGURE 8.1 Modern PEM fuel cell design [cathode: $O_2 + 4H^+ + 4e^- \rightarrow 2H_2O$, anode: $2H_2 \rightarrow 4H^+ + 4e^-$, overall: $2H_2 + O_2 \rightarrow 2H_2O$]. (Reproduced with permission from Stanford University.)

such as turning a motor) and return to the cathode side of the fuel cell. Meanwhile, on the cathode side of the fuel cell, oxygen gas (O_2) is being forced through the catalyst, where it forms two oxygen atoms. Each of these atoms has a strong negative charge. This negative charge attracts the two H^+ ions through the membrane, where they combine with an oxygen atom and two of the electrons from the external circuit to form a water molecule (H_2O). All these reactions occur in a so-called *cell stack*. The design also involves the setup of a complete system around the core component that is the cell stack. The stack will be embedded in a module including fuel, water and air management, and coolant control hardware and software. This module will then be integrated into a complete system to be used in different applications. Due to the high energetic content of hydrogen and high efficiency of fuel cells (55%), this great technology can be used in many applications like transport (cars, buses, forklifts, etc.) and backup power to produce electricity during a failure of the electricity grid.

As noted earlier, Toyota Corporation has led the development of PEM hybrid fuel cells for automobiles. "Mirai" means *the future* in Japanese, and is the name is given to Toyota's new vehicle design, which is a regular mid-size, four-door sedan. Toyota's new vehicle uses fuel cell system technology, giving it an extremely advanced powertrain compared to standard vehicles in the current time. This design is the world's first vehicle design powered by a fuel cell stack unit with zero powertrain emissions. The type of fuel cell used in this model is a PEM (Figure 8.2).

The schematic of the fuel cell individual module is illustrated above. The engine bay is a 152 bhp electric motor and the Mirai model is a front-wheel drive via single speed gearing. Behind it are two high-pressure hydrogen tanks (one for storage and one for expansion) and a high-voltage nickel metal hydride drive battery. The Mirai is capable of regenerating and storing energy under the braking like any other standard Toyota vehicle. On a full tank of hydrogen, the vehicle can compete with a similar engine/tank-sized petrol car. Refueling the vehicle will take around 3–5 minutes, and after that vehicle will be tested and proven for a strong performance, smooth drive, and quiet car with no tailpipe emissions other than water vapor. The vehicle weight fits within the range of standard sedan vehicles such as Ford Mondeos and Toyota Avensis. Even with the advanced power train technology the total weight of the vehicle is 1.8 tons, which is approaching what a beam axle can be expected to suspend without any compromise to either ride or handling [4]. The hydrogen tank is the main

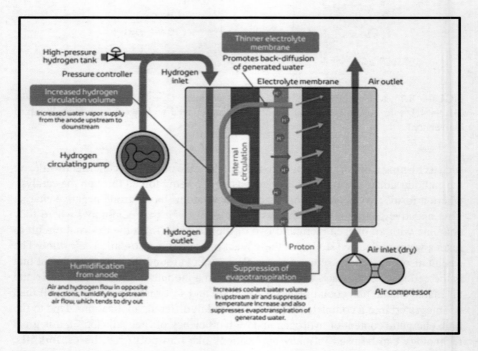

FIGURE 8.2 Toyota Mirai PEM fuel cell. (Courtesy Toyota-Japan.)

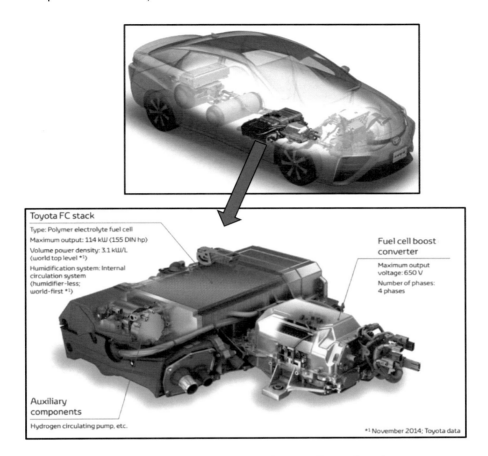

FIGURE 8.3 PEM fuel cell in the Toyota Mirai. (Courtesy Toyota-Japan.)

issue when designing fuel cell units for vehicles, but Toyota developed its own wove carbon fiber tank which contains the hydrogen to fuel the fuel cell unit. The size of the fuel cell itself is similar to conventional petrol tank and it sits under the front seats, and the fuel cell has the ability to work at cold starts as low as -30°C. The positioning of the hydrogen tank and fuel cell module is shown in Figure 8.3.

Furthermore, the cells operating principles are outlined in a step-by-step procedure, as demonstrated in Figure 8.4. The main components required to make the Mirai a functioning fuel cell–electric vehicle are displayed in Figure 8.5. Overall, the main problem faced when designing a fuel cell is the sizing and pressurizing the hydrogen tank. Various companies have solved this problem and patented designs. However, as shown in the fuel cell, designs that currently exist are used in relatively larger application bodies compared to vehicles. There are not many automotive vehicles manufacturers who have successfully patented fuel cell-powered cars. Companies such as Ford are currently in the process of completing a fuel cell drive system. The current

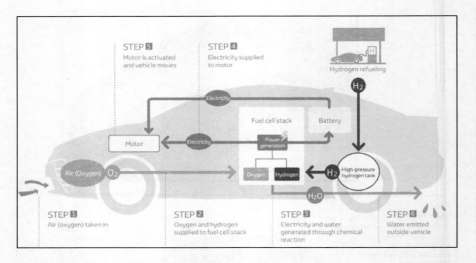

FIGURE 8.4 Fuel cell operation stages (Toyota, 2014).

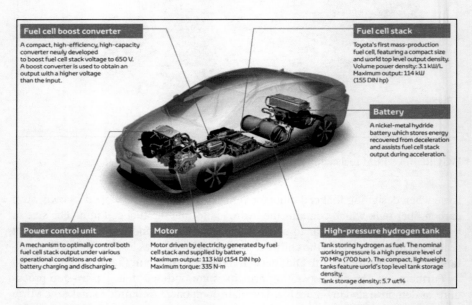

FIGURE 8.5 PEM fuel cell integrated engineering in the Toyota Mirai (Toyota, 2014).

existing design by Ford is for the transit van, as it is a large body, and all components can fit. Japanese engineering is known worldwide, and Toyota being the global lead vehicle manufacturing company has developed a compact sedan-sized vehicle which is powered by a fuel cell. This has set the blueprint for all other corporations in the 21st century.

An important contribution to the fuel cell revolution has been the implementation of computational fluid dynamics (CFD) simulation. CFD [5] provides an inexpensive and powerful method for predicting many aspects of fuel performance. In conjunction with experimental testing, it is the leading tool used in engineering fluid dynamics in the 21st century. In the automotive sector, many different codes are used. Fuel cell technology has various modeling methods since it features both structural and fluid flow phenomena. There are various simulation tools that are capable of landscaping fuel cell geometries and achieving theoretical predictions of fuel cell performance. Simulation packages are useful and cost/time effective tools that require an input, in return giving an output that satisfies the objective. Some examples of automotive-applied CFD packages include AVL Fire which is capable of simulating fluid dynamic problems that involve complex geometries and advanced physics and/or chemistry. The electrified technology is a programming language that is a multi-purpose thermo-fluid CFD software package, which defines both the powertrain and the automotive industry. Multiphase flow modules contain the Lagrangian and Eulerian multiphase modules. The Lagrangian multiphase modules entail the droplet break-up turbulence dispersion, coalescence, collision, drag, evaporation, and distortion as well as the droplet–wall interaction. The Eulerian multiphase modules consider modeling the multi-fluid of the multiphase flow, thus solving the calculation of the volume fraction distribution for all flow variables in addition to each phase. Another code is CFD-ACE+ that provides coupled simulations of thermal, fluid, chemical phenomena. The software is designed for high-performance workstations and clusters for parallel computing. However, the package is also utilized on normal computers. The software package entails built-in electrochemical models which can be further developed, i.e. manipulating the flow through porous media and small channels. Fuel cell modules account for the fundamental physics of fuel cells, and the package includes a model for water transport via the membrane. The package includes a sample model for transport of liquid water saturation through a porous media and the liquid saturation example model for multi-phase flow in the channels. COMSOL Multiphysics is adaptable programming platform for various coupled phenomena and uses finite element methods. The underlying electrochemical phenomena can be modeled using the electrolytes and electrodes provided by the fuel cell and battery modules. The CFD module gives an understanding of flow in porous media as well as multiphase flows. The model is created using the multiphase mixture, Eulerian–Eulerian multiphase models, or the bubbly flow. To highlight and describe the phase changes, the built-in step functions are utilized. The most popular software for commercial CFD remains ANSYS FLUENT which is a versatile but general-purpose fluid analysis simulation package, with a wide range of physical modeling capabilities for modeling turbulence heat transfer, flow, and reactions for industrial applications. The electrochemistry, mass and current transportation, liquid water formation, and heat source can be modeled using this software as the fuel cell module is provided as a built-in feature. To calculate the multiphase flow, two distinct approaches are utilized – Eulerian–Eulerian and Eulerian–Lagrangian. When using the Eulerian–Eulerian method the various phases are treated mathematically as inter-penetrating continua. In the fluent version of the ANSYS software there are three Eulerian–Eulerian multiphase models available: the mixture model, VOF model, and the Eulerian model. On the contrary, in the Eulerian–Lagrangian approach the multi-phase fluid is treated as a continuum by solving Navier–Stokes equations, although the dispersed phase is solved by tracing a large number of bubbles, particles, or droplets, which is attained by calculating the flow field. Finally, Open FOAM is a diverse open-source CFD code that can solve various problems from solid dynamics and electromagnetics to complex fluid flow which include chemical reactions, heat transfer, and turbulence. The object-orientated design of the programming language was written in C++, which authorizes the user to develop and implement their own numerical algorithms and models. The program is known for its adaptability as the user has complete freedom to customize and extend all existing functionalities. The approach to solving multiphase flows ranges from a system with one phase dispersed to two fluid-phase models via VOF phase fraction-based interface capturing approach, to multi-fluid models and multiphase mixture. Macedo-Valencia et al. [6] employed ANSYS FLUENT to simulate single-phase, three-dimensional fluid flow, heat transfer, electrochemical reaction, and species transport in a PEM fuel cell stack with five single cells including the membrane, gas diffusion layers, catalyst layers, flow channels, and current collectors. They observed that the species concentration is invariably higher at inlets and is reduced gradually along the channels. Furthermore, they found that minimal temperature arises at the inlet of the cathode where oxygen is supplied at temperature of 300 K. Likewise, the heat sources in PEM fuel cells are strongly interdependent on the current density distributions through membrane electrodes assembly. Awotwe et al. [7] employed ANSYS CFX to study the effect of varying the flow rate (i.e. velocity) on the pressure drop in bi-polar plate design of PEM fuel cells. They noted that a reduction in pressure drop contributes to the improvement of performance; however, many other electrochemical and geometric factors also influence overall performance of the cell, and reduction in pressure drop alone is not sufficient. In addition to these codes, many other approaches have been adopted for PEM fuel cell CFD simulation. For example, Ravishankar et al. [8] presented three-dimensional numerical computations of cooling channel designs based on traditional serpentine and spiral designs for a PEM fuel cell, using a streamline upwind/Petrov Galerkin finite element method for Reynolds number ranging from 415 to 1,247. They observed that hybrid designs achieve improved performance compared to serpentine geometries in terms of uniformity in temperature distribution at all Reynolds numbers. Many other approaches have been explored including molecular dynamics simulation, although it is computationally very expensive [9]. Other interesting studies of CFD modeling of PEM fuel cells are reported in [10–20]. In the current presentation, the ANSYS FLUENT workbench software is deployed to simulate flow characteristics and temperature distributions in a two-dimensional enclosure replicating a hybrid PEM hydrogen/oxygen fuel cell. Extensive visualization of transport phenomena in the fuel cell is included, i.e. streamline and isotherm contours in addition to density distribution. Validation of the finite volume computations has also been achieved with a thermal Lattice Boltzmann method (LBM) [21] achieving excellent agreement. Mesh independence tests are also performed. The simulations constitute a first step in understanding more deeply the intrinsic transport convection characteristics of PEM hydrogen/oxygen fuel cells.

8.2 ANSYS FLUENT CFD MODEL

To simulate a full PEM fuel cell is very challenging and requires usually of the order of millions of elements (finite volumes). It is common in commercial groups to simulate in three dimensions the PEM fuel cell and compute velocity and pressure drops with multi-million density meshes. In this presentation, however, owing to mesh limitations, attention is restricted to a two-dimensional model of a moving membrane PEM fuel cell. Simulations are executed in ANSYS FLUENT version 19 (Figure 8.6).

The focus is on laminar, viscous convection flow to compute heat flux variation, change in streamline (iso-velocity), pressure, and temperature for five various fuel cells in a layer of the fuel cell which has a moving membrane and both hydrogen and oxygen are introduced to the cell. Electrochemistry is ignored in these simulations and is the subject of a subsequent study [22]. The mathematical model employed in ANSYS FLUENT is next elaborated, followed by simulations.

8.2.1 Mass Conservation Equation

The unsteady equation for mass conservation or continuity is written as follows:

$$\frac{\partial \rho}{\partial t} + \nabla \cdot (\rho \vec{v}) = S_m \tag{8.1}$$

The equation above is the general form of the mass conservation equation which is valid for both compressible and incompressible flows. The source S_m is the mass

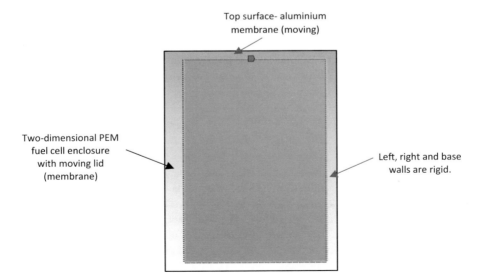

FIGURE 8.6 Model for PEM fuel cell with membrane upper wall.

(kg) applied to the continuous phase from the dispersed second phase and any user-defined sources. Here $(\vec{v})$ is the velocity vector in three dimensions, ρ is the fluid density (kg/m³), and t is the time (s).

8.2.2 Momentum Conservation Equation

Conservation of momentum in an inertial (non-accelerating) reference frame takes the form:

$$\frac{\partial}{\partial t}(\rho \vec{v}) + \nabla \cdot (\rho \vec{v} \vec{v}) = -\nabla p + \nabla \cdot (\bar{\bar{\tau}}) + \rho \vec{g} + \vec{F} \tag{8.3}$$

where, p is the static pressure (Pa), $\bar{\bar{\tau}}$ is the stress tensor (N/m²), and $\rho \vec{g}$ and $\vec{F}$ are the gravitational body force (N) and external body force (N) (e.g. thermal buoyancy), respectively arising from interaction with the dispersed phase. $\vec{F}$ also has other model-dependent source terms such as user-defined sources and porous media. The $\bar{\bar{\tau}}$ term is given by:

$$\bar{\bar{\tau}} = \mu \left[\left(\nabla \vec{v} + \nabla \vec{v}^T \right) - \frac{2}{3} \nabla \cdot \vec{v} I \right] \tag{8.4}$$

Here μ is the dynamic viscosity (kg/(ms)), I is the unit tensor and the last term shown is the effect of volume dilation. No slip velocity conditions are applied at the rigid walls and a moving boundary velocity at the top wall.

To simulate thermal convection heat transfer in the PEM fuel cell, ANSYS permits applying the heat transfer function within the fluid/solid body of the model, where problems can range from thermal mixing within a fluid to conduction in composites. To apply heat transfer to the actual model, thermal boundary conditions should be supplied to the body and material properties should be inserted that govern heat transfer. ANSYS FLUENT solves the energy transport (heat) using the general equation:

$$\frac{\partial}{\partial t}(\rho E) + \nabla \cdot (\vec{v}(\rho E + p)) = \nabla \cdot \left[k_{\text{eff}} \nabla T - \sum_j h_j \vec{J}_j + (\bar{\bar{\tau}}_{\text{eff}} \cdot \vec{v}) \right] + S_h \tag{8.5}$$

Here k_{eff} is the effective conductivity ($k + k_t$, where k_t is the thermal conductivity). $\vec{J}_j$ is the diffusion of flux species j. S_h is the heat produced due to the chemical reaction and any form of volumetric heat source. The remainder terms on the right side of the equation show the energy transfer due to species conduction, species diffusion, and viscous dissipation, respectively. Of course when electrochemistry is neglected the flux species term is neglected. The energy transport equation in solid regions (thermal conduction), within ANSYS FLUENT has the following form:

$$\frac{\partial}{\partial t}(\rho h) + \nabla \Delta (\vec{v} \rho h) = \nabla \Delta (k \nabla T) + S_h \tag{8.6}$$

Computational Fluid Dynamic Simulation

Here k is material (wall) conductivity, T is the temperature of the body (K), S_h is the volumetric heat source, ρ is the density of material (kg/m³) and h is enthalpy (J/kg). In this study, we consider the thermal convection flow in a 2D PEM fuel cell. This allows the simulation of general internal circulation in the fuel cell. Although the 2D mesh does not consider the third dimension, it enables a comprehensive understanding of the behavior of the two gases in the fuel cell membrane (oxygen and hydrogen). Heat is added to the bottom wall of the fuel cell as the top wall is a moving membrane to illustrate the oxygen and hydrogen flow valves. The bottom wall is heated to mirror the heat energy dissipated to surroundings once the hydrogen and oxygen molecules enter the center chamber (membrane). The opposite reaction to the heat dissipation is the actual heat which is supplied to the fuel cell wall itself. This analysis monitors the behavior of hydrogen and oxygen within the membrane when heat is applied. The parameters analyzed are as follows:

i. Heat flux
ii. Temperature
iii. Pressure
iv. Streamline velocity

The parameters prescribed in the model are extensive and summarized in Table 8.1. Boundary conditions prescribed are listed in Table 8.2.

To apply the ANSYS theory and determine the behavior of the two gases within the fuel cell, ANSYS FLUENT was utilized. The CFD solver can be run in or outside the workbench environment (Table 8.3).

To begin the design procedure, the parameters and dimensions of the fuel cell model were researched. Although the patented models in existence do not provide

TABLE 8.1
Data Specification in ANSYS FLUENT

Material	Hydrogen	Oxygen
Density ρ (kg/m³)	1.2999	0.08189
Specific heat C_p (J/kg-K)	Piecewise-polynomial	Piecewise-polynomial
Thermal conductivity (w/m-K)	0.0246	0.1672
Dynamic viscosity μ (kg/m/s)	1.919×10^{-5}	8.411×10^{-6}
Molecular weight (kg/mol)	31.9988	2.01594
Standard state enthalpy (J/kgmol)	0	0
Standard state enthalpy (J/kgmol-K)	205,026.9	130,579.1
Reference temperature (K)	298.15	298.15
Thermal accommodation coefficient	0.9137	0.9137
Momentum accommodation coefficient	0.9137	0.9137
Critical temperature (K)	154.58	32.98
Critical pressure (Pa)	5,043,000	1,293,000
Critical specific volume (m³/kg)	0.002293	0.031846
Acentric factor	0.021	−0.217

TABLE 8.2
Velocity and Thermal Boundary Conditions

Boundary Condition	(Temperature K)	Velocity
Moving membrane (Top wall)	-	1 m/s
Left wall	300	(No slip)
Right wall	300	(No slip)
Bottom wall	353, 473, and 1,273	(No slip)

TABLE 8.3
Moving Membrane (Top Wall) Material Properties

Membrane Wall Properties	
Length (m)	0.3
Width (m)	0.2
Material	Aluminum
Density of aluminum (kg/m^3)	2,700
Mass of membrane (kg)	0.162

dimensions of the fuel cell as the information is classified, dimensions were attained by analyzing the stack fixture within an actual vehicle. The dimensions were then researched upon, which showed similarity between the predicted dimensions measured from the stack geometry thus, giving the fuel cell membrane the dimensional image, it has. The modeling procedure was conducted in four stages: geometry, mesh, setup (pre-processing), solution (processing), and results (post-processing). A 2D x-y geometric model is designed (Figure 8.7). Once the membrane shape was outlined, the dimensions were specified as shown. The next step was to create a surface body in the main toolbox and select surfaces from sketches.

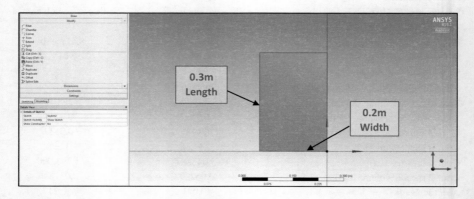

FIGURE 8.7 ANSYS PEM fuel cell geometric model.

Computational Fluid Dynamic Simulation

The mesh is initially designed using a coarse density with quadrilateral elements. It is then refined to achieve mesh (grid) independence. 10,100 elements are finally utilized in the mesh, as shown in Table 8.4 and Figure 8.8.

In the analysis "setup", the following are specified: processing options – serial, display options – enable both display mesh after reading and workbench color scheme and options – ensure double precision is disabled and dimensions is pre-selected on 2D. The pressure-based solver is selected (since incompressible flow is considered in the PEM fuel cell), (planar) stead state time response is adopted, and the number of iterations is selected as 2,000 to ensure convergence and accuracy. In the general setup, viscous laminar, single-phase flow with energy is selected ("radiation" and "multiphase" options are deselected). Within the general setup window, the fuel cell model is meshed in a steady-state time response in planar spatial dimensions. The velocity formulation option selected is the "absolute velocity" as the gases within the membrane are monitored, whereas, the relative velocity option shows the moving

TABLE 8.4
Mesh Selection Study

Number of Contours	Number of Divisions	Number of Elements
100	100	10,100

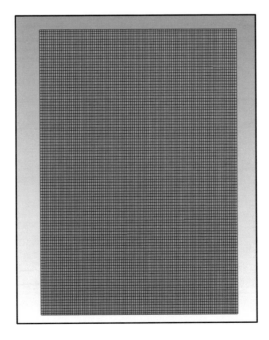

FIGURE 8.8 PEM fuel cell mesh.

membrane (top wall) which effects the behavior of the gas and describes inaccurate results. If electric fields were applied, the relative velocity would be the parameter to be analyzed. However, this is not considered here. The hydrogen and oxygen entering is not "compressed" so the pressure-based model (solver) is appropriate. The next step is to add "material" to the membrane body. The first simulation for the fuel cell membrane is with hydrogen and the subsequent analysis is conducted with oxygen. ANSYS FLUENT has its own database which enables the user to choose the material. The material properties such as density ρ (kg/m^3), specific heat C_p (J/kgK), thermal conductivity (W/mK), and dynamic viscosity μ (kg/m/s), are automatically adjusted once the material is chosen from the database. The next stage involves applying boundary conditions (Table 8.2). An initial boundary condition was applied to the moving membrane (top wall). Similarly, the same steps were followed ensuring all other surfaces (left, right, base walls) of the body have a no-slip boundary condition. Furthermore, the right and left wall as well as the moving membrane have a thermal boundary condition of room temperature at approximately 300 K. However, the base walls temperature is varied in the simulations between 80°C (353 K) and 200°C (473 K) and finally 1,000°C (1,273 K). The bottom wall was treated as a heat source due to the thermal energy dissipating from the fuel cell body when the current is extracted. The temperatures represent the thermal boundary condition at which the PEM, SOFC (solid oxide fuel cells) and AFC/PAFC (alkaline/phosphoric acid) fuel cells modules work when powering any device/s. The solver stages are summarized in Figure 8.9a and b for the "methods" and "controls" selected.

As the calculation progresses the surface monitor history was plotted as shown in the graphics window above. The solution was automatically stopped by ANSYS FLUENT once the residuals reached their specified value or after 2,000 iterations. The number of iterations varies according to the platform utilized (Figure 8.10).

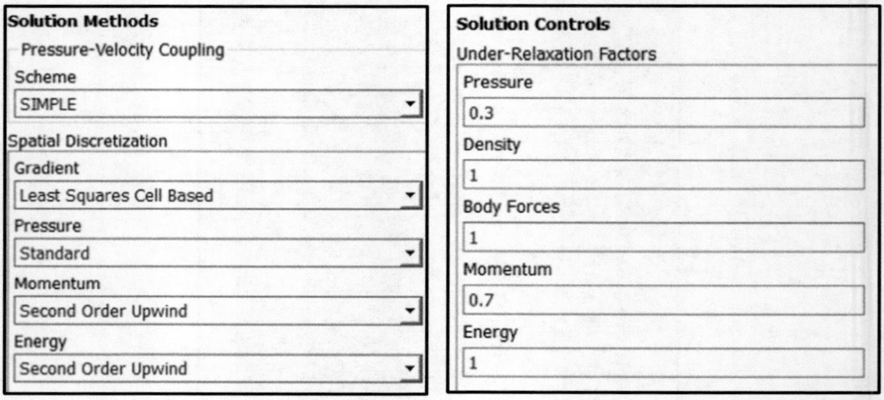

FIGURE 8.9 (a) ANSYS FLUENT solution method selection. (b) ANSYS FLUENT solution control selection.

Computational Fluid Dynamic Simulation 147

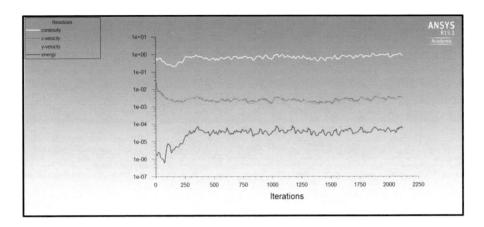

FIGURE 8.10   Residual iterations.

8.3   CONVERGENCE STUDY

Extensive simulations have been conducted in ANSYS workbench. However, to ensure that the computations are accurate, a convergence and mesh independence study is carried out to determine the element size suitable for this model. We therefore describe the outlines when the results converge, and the division size chosen to highlight the model parameters and specifications. Mesh convergence is related to the size of the element (how small the element must be) to ensure the results produced from the CFD analysis, are not affected by changing mesh size. It is one of the key issues studied in computational fluid (and solid) mechanics, to analyze the affects it has on the accuracy of the results. A convergence study can be completed for both 2D and 3D problems, and the mesh is doubled each step, so an accurate comparison can be made. Convergence studies can be carried out on velocity, temperature, heat flux, pressure, etc. A minimum of three points need to be considered, and as the mesh density increases, the variable under consideration starts to converge to a specific value. If two subsequent mesh refinements do not alter the results substantially, then it can be assumed the results have converged and grid independence is said to be achieved (Figure 8.11; Table 8.5).

In all the subsequent computations, by selecting the contour and streamline options the image of the hydrogen/oxygen behavior is portrayed, and the red-blue color highlights show the intensity of the parameter being analyzed. In this case, the three parameters analyzed were temperature, pressure, and streamline velocity. The parameter was changed from the variable section and after choosing a parameter like density, the "apply cell" option is selected. This produces very clear and coherent visualization as shown in Figure 8.12.

8.4   NUMERICAL RESULTS, VISUALIZATION, AND DISCUSSION

Two gases are simulated within the fuel cell membrane: oxygen and hydrogen. Temperature, pressure, and streamline (iso-velocity) contours are computed in

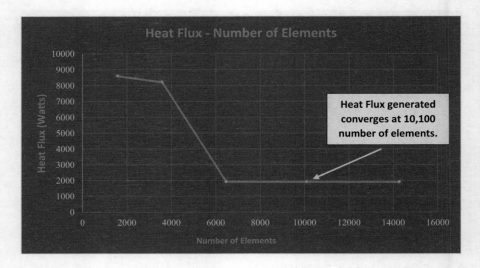

FIGURE 8.11 Mesh independence graph.

TABLE 8.5
Mesh Independence Study

Number of Divisions	Number of Elements	Heat Flux at 0.1 m
40	1,599	8,591.78418
60	3,600	8,211.40723
80	6,480	1,919.6947
100	10,100	1,909.78406
120	14,280	1,899.80591

addition to density distribution. The results are displayed below for both gases (hydrogen and oxygen) at various temperatures (353, 473, and 1,273 K). The temperatures illustrate the working temperature of the following fuel cell types:

i. PEM fuel cell – operation temperature – 353 K (80°C)
ii. AFC and PAFC fuel cell – operating temperature 473 K (200°C)
iii. SOFC – operating fuel cell – 1,273 K (1,000°C)

Based on the mesh designed earlier, the ensuing plots depict temperature, pressure, and streamline sets for the three different fuel cell configurations studied, i.e. PEM

Computational Fluid Dynamic Simulation 149

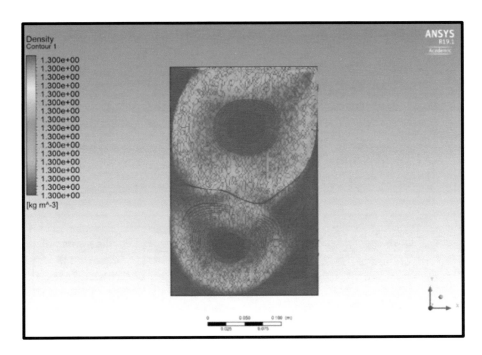

FIGURE 8.12 Density distribution computation for simulations.

fuel cell, AFC and PAFC fuel cell, and SOFC. In all the distributions, the common pattern highlighted by the results shows *asymmetrical dual vortices* with intense streamlines around the center points for both gases. The buoyancy effect (natural convection) occurs in the results shown by pressure plots for both oxygen and hydrogen gas for all three types of fuel cells. The buoyancy effect is an upward force exerted by the fluid which opposes the weight of the immersed gas. In a column of fluid pressure increases with the depth owing to the weight of the overlying fluid (air). Thus, the pressure at the bottom of the column is greater than the top of the column. A high Nusselt number (ratio of convective to conductive heat transfer across the boundary) arises for the PEM and AFC/PAFC fuel cells in the temperature plots. In this study convection includes both advection and thermal diffusion. Additionally, thermal properties, such as thermal conductivity, specific heat capacity and heat flux, moves differently through different gases. This explains the deviation in flow behavior of both gases shown in the contour plots. As noted earlier, the top, right and left walls are at room temperature 300 K, and the top wall, i.e. moving membrane has a velocity of 1 m/s representing the entry section.

Case 1: PEM Fuel Cell

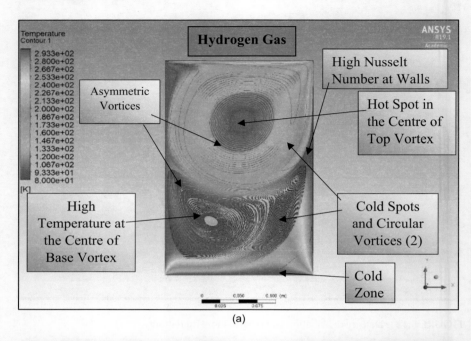

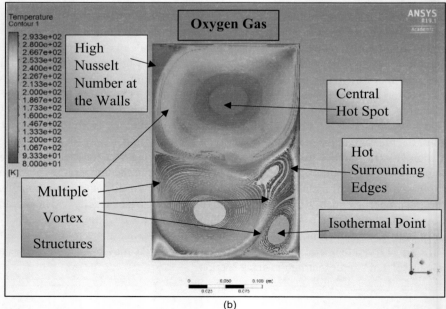

FIGURE 8.13 (a) Temperature plot (hydrogen). (b) Temperature plot (oxygen).

Computational Fluid Dynamic Simulation

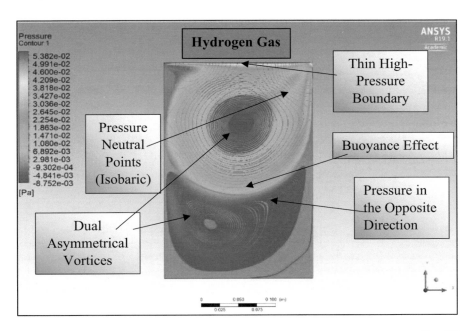

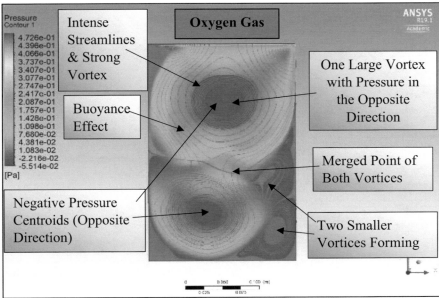

FIGURE 8.14 (a) Pressure contour distribution (hydrogen). (b) Pressure contour distribution (oxygen).

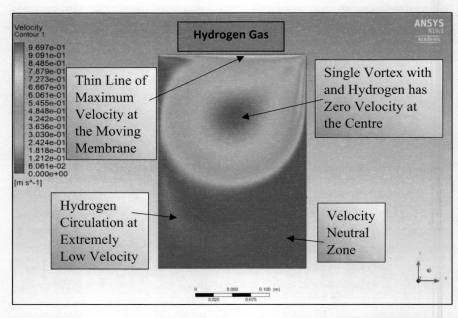

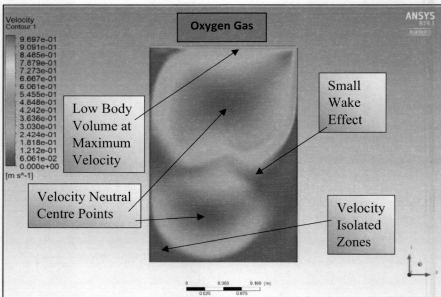

FIGURE 8.15 (a) Streamline contour distribution (hydrogen). (b) Streamline contour distribution (oxygen).

Computational Fluid Dynamic Simulation

Case II: AFC/PAFC Fuel Cell

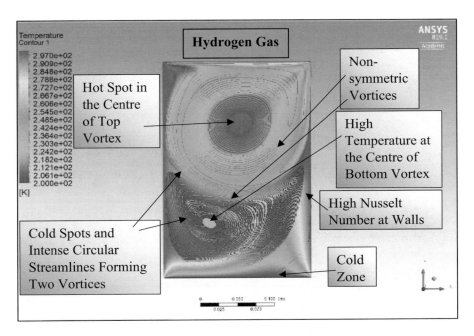

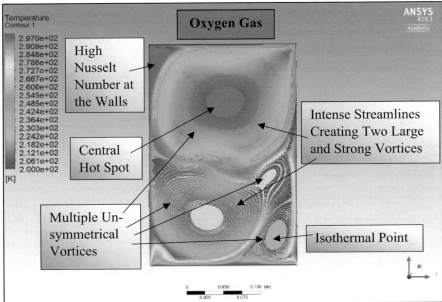

FIGURE 8.16 (a) Temperature contour distribution (hydrogen). (b) Temperature contour distribution (oxygen).

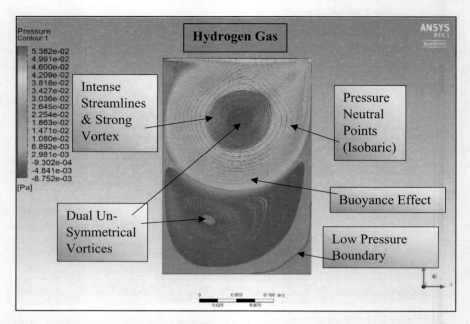

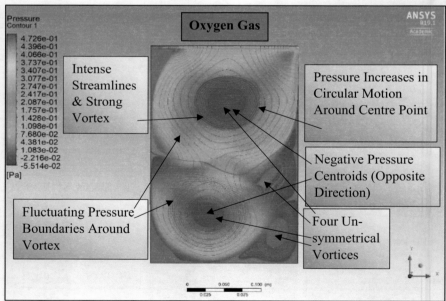

FIGURE 8.17 (a) Pressure contour distribution (hydrogen). (b) Pressure contour distribution (oxygen).

Computational Fluid Dynamic Simulation 155

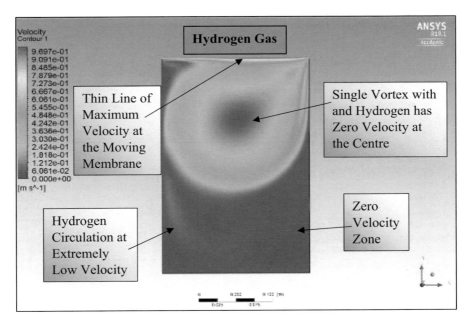

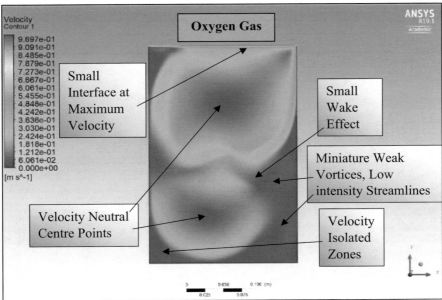

FIGURE 8.18 (a) Streamline **(iso-velocity)** contour distribution (hydrogen). (b) Streamline **(iso-vel**ocity) contour distribution (oxygen).

Case III: SOFC Fuel Cell

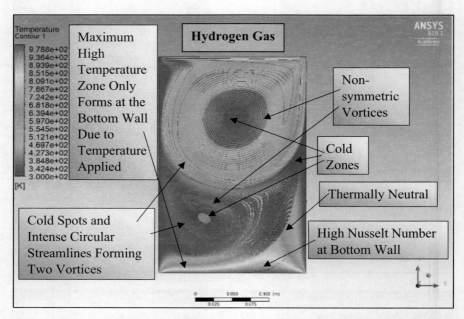

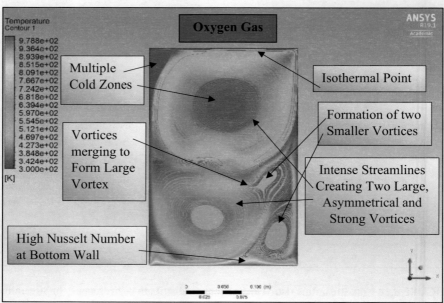

FIGURE 8.19 (a) Temperature contour distribution (hydrogen). (b) Temperature contour distribution (oxygen).

Computational Fluid Dynamic Simulation 157

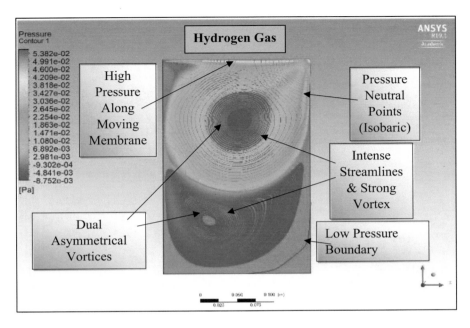

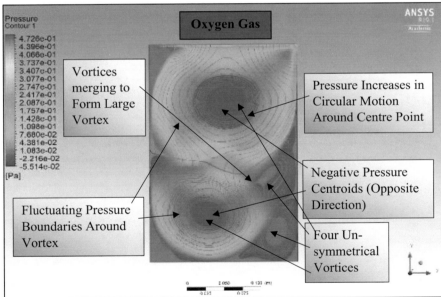

FIGURE 8.20 (a) Pressure contour distribution (hydrogen). (b) Pressure contour distribution (oxygen).

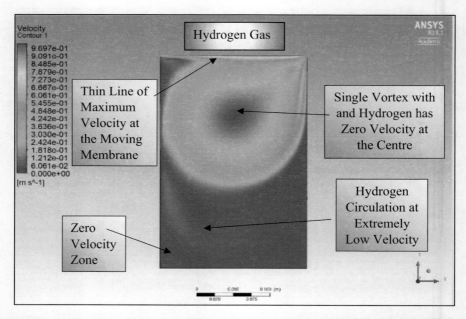

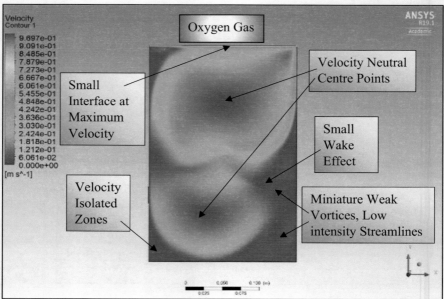

FIGURE 8.21 (a) Streamline (**iso-velocity**) contour distribution (hydrogen). (b) Streamline (**iso-velocity**) **contour** distribution (hydrogen).

Computational Fluid Dynamic Simulation

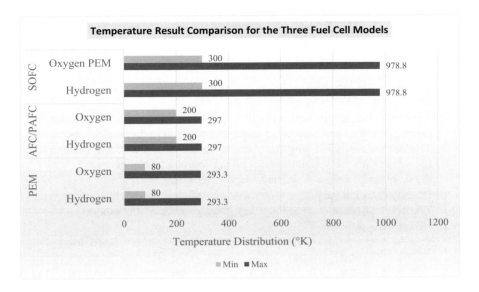

FIGURE 8.22 Temperature distribution comparison between all three models.

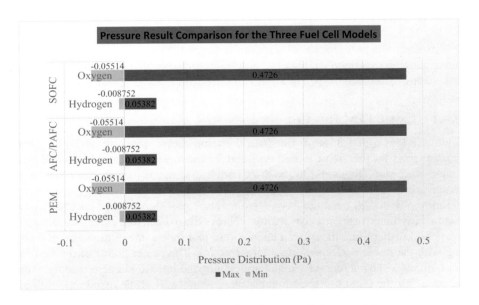

FIGURE 8.23 Pressure distribution comparison between all three models.

Furthermore, Figures 8.22 and 8.23 summarize the temperature and pressure results for all three fuel cells. Implicit in the simulation have been the following model assumptions:

i. The gases are not ideal.
ii. The stack is fed with hydrogen and air.
iii. The stack is equipped with a cooling system which maintains the temperature at the anode and cathode exits stable and equal to the stack temperature.
iv. The stack is equipped with water management system ensuring the humidity within the cell is maintained at an appropriate level at any load.
v. Pressure drops across flow channels are negligible.
vi. Density does not vary.
vii. The cell resistance is constant at any given condition of operation.
viii. No radiation effects.
ix. Prandtl number (ratio of momentum diffusivity to thermal diffusivity) is constant for each scenario.
x. Wall temperatures (left, right, and top are constant at room temperature).
xi. The top wall has a velocity of 1 m/s as it represents the moving membrane (gas entry).
xii. The rate at which the bottom wall temperature increases is constant (manually changed).
xiii. Model is mesh dependent.
xiv. Model is isotropic.

It is important to note that the distinctions between the fuel cells relate to the membrane material (e.g. polymer, alkaline, and solid oxide) but not to the fuel gases which are either hydrogen or oxygen in all three sets of simulations considered.

8.4.1 Case I: PEM Fuel Cell

Figures 8.13a, b and 8.15a, b illustrate the temperature, pressure, and streamline contour plots for this first fuel cell type and for hydrogen fuel and oxygen fuel. The temperature plots (Figure 8.13a and b) for both hydrogen and oxygen gases show high temperatures at the wall implying a high mean Nusselt number. The hydrogen temperature plot differs from the oxygen temperature plot shown since it exhibits only has two asymmetrical vortex structures whereas the oxygen plot generates four such vortices. Furthermore, the oxygen temperature plot shows one of the vortices merging with the lower vortex to form a larger vortex. The center points of the vortices for both plots show a relatively high temperature and intense low-temperature zones surrounding the point itself. Commonly, the bottom wall of the membrane shows a cold zone when both hydrogen and oxygen are inserted on the membrane platform. The maximum temperatures at the surrounding walls for either hydrogen or oxygen cases are 293.3 K. The coldest areas within the membrane are at a temperature of 80 K. This is due to the convection that occurs between the gas and membrane body which leads to redistribution of heat in the fuel cell enclosure. The pressure plots for the hydrogen and oxygen gases within the membrane are shown in Figure 8.14a and b. A strong influence from thermal buoyancy force generates the formation of higher pressure in the opposite direction in the vicinity of the lower vortices (cells). Similar to the temperature plots the two vortices are again asymmetrical. There is a high-pressure zone shown at the top wall due to the moving membrane and high-pressure center points between the vortices in the opposite direction for both oxygen and hydrogen gas. The difference between the hydrogen and oxygen gas within the membrane is the formation of two small vortices when oxygen is used. Using hydrogen shows a single streamline at the bottom left chamfering the PEM fuel cell wall; however, this further develops when oxygen is used, and the intensity and volume of the streamlines increases, manifesting in the synthesis of two supplementary circulation zones, i.e. cells. The pressure shown by one of the small vortices is substantially increasing and eventually merging with the large lower vortex is observed. The maximum pressure zones for when hydrogen and oxygen are inserted on the membrane are 5.382×10^{-2} and 4.73×10^{-1} Pa, respectively. The low-pressure areas and neutral areas are shown in green and the negative pressures represent pressure in the opposite direction. Figure 8.14a illustrates the lower base vortex with a pressure in the opposite direction at 8.75×10^{-3} Pa when hydrogen is inserted into the membrane. Figure 8.14b shows the bottom vortex with a pressure in the opposite direction at 5.51×10^{-2} Pa when oxygen is inserted into the membrane. This is again intimately associated with the thermal buoyancy effect (relative to the viscous hydrodynamic force) which exacerbates thermal convection between the fuel gas and air in the membrane body. Finally, Figure 8.15a and b indicates that the streamline velocity plots exhibit the largest dissimilarity compared to the temperature and pressure plots when the gas changes from hydrogen to oxygen. When hydrogen is inserted on the membrane platform there is one vortex which forms at the top of the fuel cell membrane. The vortex has a velocity neutral center point and circulation is constrained with gas movement is limited at the upper region of the fuel cell. There is a high-velocity zone at the moving membrane which is due to the applied velocity at the top wall of 1 m/s, i.e. the moving lid boundary condition. Therefore, the gas within the cell membrane platform has a low velocity relative to the top wall thus and progressively the gas circulates at velocities beneath 1 m/s. For the oxygen fuel case, within the PEM design, two vortices form, and the streamline intensity is relatively high, thus generating consistent gas circulation and spatially extensive low-velocity zones. The streamlines merge to form a circulation pattern observed in both hydrogen and oxygen velocity plots. There is also a wake effect illustrated which arises due to the high-velocity circulation around the upper vortex. The blue zones highlight velocity neutral zones and the gas is moving at 0 m/s at those points. Both figures show the high-velocity zone at the moving membranes' top wall with a velocity value of 0.9697 m/s. Even though the maximum velocity is 1 m/s, the highest velocity measured of the gas movement is lower, which is due to frictional factors associated with viscous hydrodynamic forces.

8.4.2 Case II: AFC/PAFC Fuel Cell

Figure 8.16a and b visualizes the temperature contour distributions for both hydrogen and oxygen fuel gas cases and it is apparent that high temperatures arise along the left wall, right wall and also top wall. The thermal distribution is therefore similar to the PEM fuel cell setup at the fuel cell boundaries. The hydrogen temperature plot (Figure 8.16a) differs from the oxygen temperature plot (Figure 8.16b) mainly in that it exhibits only two asymmetrical vortices whereas the oxygen plot has four distinct circulation zones (vertices). Furthermore, the oxygen temperature plot shows one of the vortices merging with the lower vortex to form a larger vortex. The oxygen plot also shows a relatively large cold surface area however, the hydrogen plot shows relatively colder area by the dark blue zone shown by the lower vortex. The center points of the top vortices for both plots show a relatively high temperature and intense low-temperature streamlines surrounding the point itself. Commonly, the bottom wall of the membrane shows a cold zone when both hydrogen and oxygen are inserted on the membrane platform. The maximum temperatures at the surrounding walls for when hydrogen and oxygen are inserted on the membrane are 297 K. The coldest areas within the membrane are at a temperature of 200 K. This is due to the convection that occurs between the gas and membrane body. In comparison to the PEM fuel cell there is approximately a 30% change in temperature whereas, the PEM cell shows a 60% temperature difference between the hot and cold zones. The pressure plots for the hydrogen and oxygen fuel gas cases are shown in Figure 8.17a and b. Both reveal that higher pressure zones are synthesized in the opposite direction in the vicinity of the lower vortices. The two vortices which form are asymmetrical for both hydrogen and oxygen. There is a *high-pressure zone* shown at the top wall due to the moving membrane which is greater for the hydrogen model compared to the oxygen model. Furthermore, there are highly concentrated pressure points between the vortices in the opposite direction for both oxygen and hydrogen gas. The difference between the hydrogen and oxygen gas within the membrane is the formation of two small vortices when oxygen is used, which is similar in the AFC/PAFC system geometry, but not identical to the PEM fuel cell model. The hydrogen model shows a single streamline at the bottom left chamfering the fuel cell surface however, it further develops when oxygen is used, and the intensity of the streamlines increases, producing two extra vortices. The pressure shown by one of the small vortices is substantially elevated and eventually there is a fusion with the large lower vortex. The maximum pressure zones for when hydrogen and oxygen are computed on the membrane are 5.382×10^{-2} and 4.726×10^{-1} Pa, respectively. The pressure neutral areas are shown in green and the negative pressures represent pressure in the opposite direction. Figure 8.17a shows the bottom vortex with a pressure in the opposite direction at 8.75×10^{-3} Pa when hydrogen is inserted into the membrane. Figure 8.17b shows the bottom vortex with a pressure in the opposite direction at 5.51×10^{-2} Pa when oxygen is inserted into the membrane. This is again connected to the relative influence of thermal buoyancy and viscous hydrodynamic force in the enclosure fuel cell regime. Finally, the streamline velocity plots (Figure 8.18a and b) show the largest deviation computed when the gas changes from hydrogen to oxygen. However, the AFC model behaves generally similar to the PEM model. When hydrogen is inserted on the membrane platform there is a single vortex which forms at the top of the cell membrane. The vortex has a velocity neutral center point and circulation of increasing velocity which shows the gas movement is again confined principally to the upper region of the fuel cell. In addition, there is minimal hydrogen movement at speeds of 0.303 m/s at the bottom right of the membrane body. There is a high-velocity zone at the moving membrane which was due to the applied velocity at the top wall of 1 m/s, as it is a consistent design factor for all three models. Therefore, the gas within the cell membrane platform has a low velocity relative to the top wall thus and again can only achieve circulation at velocities of less than 1 m/s. When oxygen fuel is considered in the simulation, two circulation zones are generated and the streamline intensity is relatively high thus, forming consistent gas circulation in addition to low-velocity zones. This behavior is again generally of a similar nature to that computed for the PEM model as the streamlines merge to form the circulation pattern shown in both hydrogen and oxygen iso-velocity plots. There is again high velocity and more vigorous circulation localized around the upper vortex. The blue zones highlight velocity neutral zones and the gas is moving at 0 m/s at those points. Figure 8.18a and b successfully captures the high-velocity zone at the moving membranes' top wall with a velocity value of 0.9697 m/s.

8.4.3 Case III: SOFC Fuel Cell

Unlike the PEM and AFC/PAFC models, this model shows a variation in the temperature plots when both hydrogen and oxygen are used. The temperature plots for both hydrogen and oxygen gases exhibit higher magnitudes at the bottom wall where the temperature is applied. There is a markedly greater temperature at the base wall when hydrogen is applied as depicted in Figure 8.19a in comparison to oxygen (Figure 8.19b). This is probably attributable to the greater molecular weight of oxygen (atomic mass is sixteen times greater than hydrogen). The hydrogen model shows two large asymmetrical vortices. This differs from the oxygen model as the membrane shows a formation of four distinct but variable magnitude circulation zones (vortices). Furthermore, the oxygen temperature plot shows one of the vortices merging with the lower vortex to form a large vortex and in this regard the SOFC fuel cell performs similarly to the PEM and AFC/PAFC fuel cell models. The oxygen plot also shows a relatively large cold surface area however, the hydrogen plot shows relatively colder area by the dark blue zone associated with the lower vortex. The hydrogen plot shows multiple high-temperature zones as demonstrated by the circulation at the top vortex. The center points of the top vortices for both plots show a relatively low temperature and intense high temperature (isotherms) surrounding the point itself. This trend is the opposite to that computed in the PEM and AFC fuel cell models. This is due to the substantially high temperature at the bottom wall. The plot shows relative temperature and since the base wall temperature is significantly high, the red zones are in the lower half of the fuel cell enclosure. Commonly, the base wall of the membrane [12–15] shows a hot zone, when both hydrogen and oxygen are inserted on the membrane platform. The maximum temperatures at the base wall for both hydrogen and oxygen on the membrane are 978.8 K. The coldest areas within the fuel cell are at a temperature of 300 K. The temperature difference throughout the SOFC model is approximately 60% which is similar to the PEM fuel cell model. The reason why the SOFC fuel cell works at such high temperatures is due to its application such as large appliances within large structures or powering larger hybrid vehicles and also other commercial systems (entire households). The pressure plots for the hydrogen and oxygen gases within the fuel cell are shown in Figure 8.20a and b which again highlight that thermal buoyancy contributes substantially to the formation of higher pressure in the opposite direction at the base vortices. The two vortices which form are asymmetrical for both hydrogen and oxygen. There is a high-pressure zone shown at the top wall due to the moving membrane which is greater for the hydrogen model compared to the oxygen model. Furthermore, there are highly concentrated pressure points at the center of the vortices in the opposite direction for both oxygen and hydrogen gas. The difference between the hydrogen and oxygen gas within the membrane is the formation of two small vortices when oxygen is used, which is similar to both the PEM and AFC/PAFC fuel cell models. The hydrogen model shows a single streamline at the bottom left chamfering the membrane surface. This expands and further develops when oxygen is used, and the intensity and volume of the streamlines increases, leading to the formation of two extra vortices at the base half of the fuel cell. The pressure shown by one of the small vortices is substantially increasing thus, merging with the large lower vortex (Figure 8.20a). The maximum pressure zones for when hydrogen and oxygen are inserted on the membrane are 5.382×10^{-2} and 4.726×10^{-1} Pa, respectively, which is the same as for the other two fuel cell models. The pressure neutral areas are shown in green and the negative pressures represent pressure in the opposite direction. Figure 8.20a shows the base vortex with a pressure in the opposite direction at 8.75×10^{-3} Pa when hydrogen is inserted in the fuel cell. Figure 8.20b shows the bottom vortex with a pressure in the opposite direction at 5.51×10^{-2} Pa when oxygen is inserted into the membrane. Again, thermal buoyancy is a strong contributor to this pattern and intensifies the interaction between the gas and air in the fuel cell. Finally, the streamline (iso-velocity) plots (Figure 8.21a and b) exhibit significant disparity compared to the temperature and pressure plots when the fuel gas changes from hydrogen to oxygen. However, the overall behavior of the gas velocities is similar to the PEM and AFC/PAFC models. When hydrogen is inserted on the membrane platform there is one vortex which forms at the top of the cell membrane. The vortex has a velocity neutral center point and circulation with increasing velocity, indicating the gas movement is limited at the top of the cell. In addition, there is minimal hydrogen movement at speeds of 0.484 m/s at the bottom right of the membrane body. There is a high-velocity zone at the moving membrane which is associated with the applied velocity at the top wall of 1 m/s, as it is a consistent design factor for all three models. Therefore, the gas within the cell membrane platform has a low velocity relative to the top wall. Effectively once again, gas circulation is at speeds lower than 1 m/s. When oxygen is tested within the membrane, two vortices form, and the streamline intensity is relatively high, thus forming consistent gas circulation and larger but weaker velocity zones. This behavior is similar to the PEM model as the streamlines merge to form the circulation pattern shown in both hydrogen and oxygen velocity plots. The blue zones highlight velocity neutral zones and the gas is moving at 0 m/s at those points. Both streamline plots show the high-velocity zone at the moving membranes' top wall with a velocity value of 0.9697 m/s. There is also high-velocity circulation around the upper vortex. The overall behavior of the gas velocities follows the same trend for all three fuel cell models, although there are distinct variations in selected zones for temperature (isotherm), pressure, and streamline distributions.

Furthermore, deviations are computed between the hydrogen and oxygen fuel cases. Good and efficient circulation and heat transfer are achieved in all three fuel cell configurations although further insight requires electrochemistry inclusion in the cathode/anode components which has been neglected in the current study (Figures 8.22 and 8.23).

8.5 FURTHER VALIDATION WITH THERMAL LBM CODE

Although mesh independence of the ANSYS FLUENT simulations has been conducted and described earlier, further corroboration of the accuracy of the finite volume computations requires an alternative numerical solution (in the absence of experimental verification). In this regard a *particle-based technique* is selected, i.e. LBM. This computational method is based on the microscopic particle models and mesoscopic kinetic equations. LBM builds simplified kinetic models that incorporate only the essential physics of microscopic or mesoscopic processes so that the macroscopic averaged properties obey the desired macroscopic equations. This subsequently avoids the use of the full Boltzmann equation and avoids following each particle as in molecular dynamics simulations. Although LBM is based on a particle representation, the principal focus remains in the averaged macroscopic behavior. The kinetic nature of the LBM introduces three important features that distinguish this methodology from other numerical methods. Firstly, the convection operator of the LBM in the velocity phase is linear. The inherent simple convection when combined with the collision operator allows the recovery of the nonlinear macroscopic advection through multiscale expansions. Secondly, the incompressible Navier–Stokes equations can be obtained in the nearly incompressible limit of the LBM. The pressure is calculated directly from the equation of state in contrast to satisfying Poisson's equation with velocity strains acting as sources. Thirdly, the LBM utilizes the minimum set of velocities in the phase space. Since only one or two speeds and a few moving directions are required, the transformation relating the microscopic distribution function and macroscopic quantities is greatly simplified and consists of simple arithmetic calculations. For thermal convection flows, the thermal LB model [23] is utilized which employs two distribution functions, f and g, for the flow and temperature fields, respectively. It was popularized by McNamara and Zanetti [24] who utilized it in their development of a multi-speed thermal fluid lattice Boltzmann method to solve heat transfer problems. Thermal LBM models the dynamics of fluid particles to capture macroscopic fluid quantities such as velocity, pressure, and temperature. In this approach, the fluid domain is discretized to uniform Cartesian cells. Thermal LBM has been employed in several enclosure convection flows including Taoufik et al. [25,26]. It has also been used in PEM fuel cell simulations [27–29]. The probability of finding particles within a certain range of velocities at a certain range of locations replaces tagging each particle as in the computationally intensive molecular dynamics simulation approach. In thermal LBM, each cell holds a fixed number of distribution functions, which represent the number of fluid particles moving in these discrete directions. The D2Q9 model is very popular in this regard. The density and distribution functions (f and g) are calculated by solving the Lattice Boltzmann equation, which is a special discretization of the *kinetic* Boltzmann equation. After introducing the BGK approximation, the general form of lattice Boltzmann equation with external force for a generalized enclosure thermal convection problem is as follows:

For the flow field:

$$f_i(x + c_i \Delta t, t + \Delta t) = f_i(x,t) + \frac{\Delta t}{\tau_v}\left[f_i^{eq}(x,t) - f_i(x,t)\right] + \Delta t c_i F_k \qquad (8.7)$$

For the temperature field:

$$g_i(x + c_i \Delta t, t + \Delta t) = g_i(x,t) + \frac{\Delta t}{\tau_C}\left[g_i^{eq}(x,t) - g_i(x,t)\right] \qquad (8.8)$$

Here Δt denotes lattice time step, c_i is the discrete lattice velocity in direction i, F_k is the external force in direction of lattice velocity, and τ_v and τ_C denotes the *lattice relaxation times* for the flow and temperature fields. The kinetic viscosity v and the thermal diffusivity α are defined in terms of their respective relaxation times, i.e. $v = c_s^2(\tau_v - 1/2)$ and $\alpha = c_s^2(\tau_C - 1/2)$. Note that the limitation $0.5 < \tau$ should be satisfied for both relaxation times to ensure that viscosity and thermal diffusivity are positive. Furthermore, the *local equilibrium distribution function* determines the *type* of problem being simulated. It also models the equilibrium distribution functions, which are calculated with the following equations for flow and temperature fields, respectively:

$$f_i^{eq} = w_i \rho \left[1 + \frac{c_i \cdot u}{c_s^2} + \frac{1}{2}\frac{(c_i \cdot u)^2}{c_s^4} - \frac{1}{2}\frac{u^2}{c_s^2}\right] \qquad (8.9)$$

$$g_i^{eq} = w_i T \left[1 + \frac{c_i \cdot u}{c_s^2}\right] \qquad (8.10)$$

where w_i is a weighting factor and ρ is the lattice fluid (gas) density. For natural convection, i.e. with thermal buoyancy, the Boussinesq approximation is applied.

To ensure that the thermal LBM code works in the *near-incompressible regime*, the characteristic velocity of the flow for the natural convection regime $\left(V_{natural} = \sqrt{\beta g_y \Delta T H}\right)$ must be *small* compared with the fluid speed of sound. In most simulations, the characteristic velocity is adopted as 0.1 that of sonic speed. Bounce-back boundary conditions have to be applied on all solid boundaries, which indicates that incoming boundary populations are equal to outgoing populations *after* the collision. Bounce back type boundary conditions are proven to provide more accurate numerical approximations for LBM simulations. Similarly, the temperature requires bounce-back boundary conditions (e.g. adiabatic) on different boundaries. A schematic of the D2Q9 LBM mesh is shown in Figure 8.24. Approximately 40,000 cells have been implemented to attain the desired accuracy after a cell independence test. Comparison has been conducted with several ANSYS FLUENT simulations

Computational Fluid Dynamic Simulation

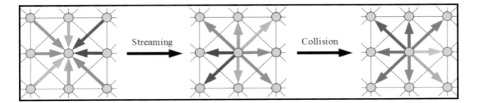

FIGURE 8.24 D2Q9 lattice in LBM.

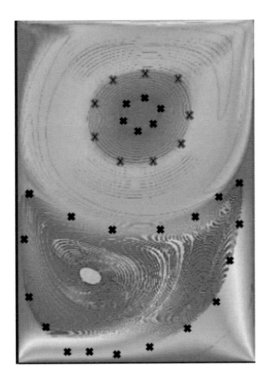

FIGURE 8.25 ANSYS FLUENT versus thermal LBM *temperature contours* (PEM hydrogen case) (N.B. crosses highlight LBM solution for outer periphery of selected circulation zones).

(*Case I: PEM fuel cell*, hydrogen gas; Figures 8.13a and 8.15a) as shown below. Excellent correlation has been attained. Compilation times were approximately 600–700 seconds on an SGI Octane desk workstation. The comparisons are shown in Figures 8.25 and 8.26 for temperature and streamlines. Confidence in the ANSYS finite volume simulations is therefore justifiably very high.

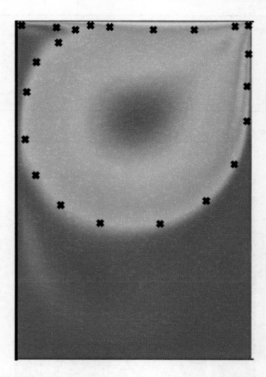

FIGURE 8.26 ANSYS FLUENT versus thermal LBM *streamline contours* (PEM hydrogen case) (N.B. crosses highlight LBM solution at outer periphery of selected circulation zone).

8.6 CONCLUSIONS

Detailed computational fluid dynamic simulations of simplified fuel cell systems are presented. ANSYS FLUENT finite volume commercial software (version 19) has been deployed to simulate flow characteristics and temperature distributions in a two-dimensional enclosure replicating a hybrid hydrogen/oxygen fuel cell of the PEM, AFC/PAFC, and SOFC types. Approximately 10,100 cells are used in the ANSYS simulations. Mesh independence tests are also performed. Extensive visualization of transport phenomena in the fuel cell is included, i.e. streamline, pressure, and isotherm contours. The top wall of the fuel cells was treated as a moving membrane which can relate to the entrance point of the gas and the bottom wall was heated which is treated as an external heat source. Validation of the finite volume computations has also been achieved with a thermal LBM with a D2Q9 grid and 40,000 cells, achieving excellent agreement. The ANSYS simulations generally demonstrate that the overall behavior of the hydrogen or oxygen fuel gas velocities is similar in the PEM, AFC/PAFC, and SOFC models. When hydrogen is inserted on the membrane platform there is one vortex which forms at the top of the cell membrane. The vortex has a velocity neutral center point and circulation with increasing velocity, indicating the gas movement is limited at the top of the cell. When oxygen is tested within the membrane, two vortices form, and the streamline intensity is relatively high thus, forming consistent gas circulation and larger but weaker velocity zones. There is also high-velocity circulation around the upper vortex. The overall behavior of the gas velocities follows the same trend for all three fuel cell models, although some distinct variations are observed in selected zones like temperature (isothermal), pressure and streamline distributions when computed for hydrogen and oxygen fuel cases. In the third configuration, i.e. SOFC, the center points of the top vortices for both plots show a relatively low temperature and intense high temperature (isotherms) surrounding the point itself. This trend is contrary to that computed in the PEM and AFC fuel cell models. This is due to the substantially high temperature at the bottom wall. The current analysis has been confined to thermal convection within a simple rectangular geometry. Further, studies have refined this geometry to consider serpentine channels, and also incorporated ANSYS FLUENT electrochemical reaction models [30] which will be reported imminently. Additionally, structural deformation can be explored to analyze the thermal stresses generated in individual fuel cell layers.

REFERENCES

1. S. Bhatt, B. Gupta, V.K. Sethi, & M. Pandey, Polymer exchange membrane (PEM) fuel cell: A review. *Int. J. Curr. Eng. Technol.*, 2(1), 219–226 (2012).
2. H. Javaid, *Investigating Fuel Cell Technology in Automobiles, BEng (Hons) Mechanical Engineering, Final Year Project Report*, Salford University, May (2019).
3. W. Vielstich, A. Lamm, & H.A. Gasteiger, *Handbook of Fuel Cells: Fundamentals, Technology and Applications*, Wiley, New York (2003).
4. https://www.toyota.co.uk/world-of-toyota/environment/fuel-cell-vehicle (2020).
5. T.J. Chung, *Computational Fluid Dynamics*, CUP, New York (2002).
6. J. Macedo-Valencia et al., 3D CFD modelling of a PEM fuel cell stack. *Int. J. Hydrogen Energy*, 41 (48), 23425–23433 (2016).
7. T.W. Awotwe, Z. El-Hassan, F.N. Khatib, A. Al Makky, J. Mooney, A. Barouaji, & A-G. Olabi, Development of bi-polar plate design of PEM fuel cell using CFD techniques. *Int. J. Hydrogen Energy*, 42(40), 25663–25685 (2017).
8. S. Ravishankar, & K.A. Prakash, Numerical studies on thermal performance of novel cooling plate designs in polymer electrolyte membrane fuel cell stacks. *Appl. Thermal Eng.*, 66, 239–251 (2014).
9. X. Li et al., Molecular dynamics simulation study of a polynorbornene-based polymer: A prediction of proton exchange membrane design and performance. *16th China Hydrogen Energy Conference (CHEC 2015)*, November, Zhenjiang City, Jiangsu Province, China (2015).
10. M. Nasri, & D. Dickinson, Thermal management of fuel cell-driven vehicles using HT-PEM and hydrogen storage. *9th International Conference on Ecological Vehicles and Renewable Energies (EVER)*, 25–27 March, Monte-Carlo, Monaco (2014).
11. A. Ranzo et al., Validation of a three dimensional PEM fuel cell CFD model using local liquid water distributions measured with neutron imaging. *Int. J. Hydrogen Energy*, 39, 7089–7099 (2014).
12. J. Becker, C. Wieser, S. Fell, & K. Steiner, A multi-scale approach to material modelling of fuel cell diffusion media. *Int. J. Heat Mass Transf.*, 54(7), 1360–1368 (2011).
13. C. Siegel, Review of computational heat and mass transfer modelling in polymer-electrolyte membrane (PEM) fuel cells. *Energy*, 33(9), 13311352 (2008).

14. M. Khan, Y. Xiao, B. Sundén, & J. Yuan, Analysis of multiphase transport phenomena in PEMFCs by incorporating microscopic model for catalyst layer structures, *Proc. 2011 ASME Int. Mech. Eng. Cong. and Exp.,* New York, IMECE2011-65142, p. 903912 (2011).
15. G. Mohan et al., Analysis of flow maldistribution of fuel and oxidant in a PEMFC, *ASME J. Energy Resour. Technol.,* 126(4), 262–270 (2004).
16. Y. Ding et al., 3D simulations of the impact of two-phase flow on PEM fuel cell performance, *Chem. Eng. Sci.,* 100, 445–455 (2013).
17. J. Cao, & N. Djilali, Numerical modeling of PEM fuel cells under partially hydrated membrane conditions, *ASME J. Energy Resour. Technol.,* 127(1), 26–36 (2005).
18. G. Besagni et al., Application of an integrated lumped parameter-CFD approach to evaluate the ejector-driven anode recirculation in a PEM fuel cell system, *Appl. Thermal Eng.,* 121, 628–651 (2017).
19. J. Wang, Simulation of PEM fuel cells by OpenFOAM. *2014 MVK160 Heat and Mass Transport Conference,* May 15, Lund, Sweden (2014).
20. Y. Xiao, Dou, M., Yuan, J., Hou, M., Song, W., & B. Sundén, Fabrication process simulation of a PEM fuel cell catalyst layer and its microscopic structure characteristics. *J. Electrochem. Soc.,* 159(3), B308–B314 (2012).
21. S. Chen, & G.D. Doolen, Lattice Boltzmann method for fluid flows. *Ann. Rev. Fluid Mech.,* 30, 329–364 (1998).
22. O. Anwar Bég, T.A. Bég, H.J. Leonard, W.S. Jouri, & A. Kadir, Electrochemical computation in hydrogen-based serpentine PEM fuel cells with ANSYS FLUENT. *J. Eng. Thermophys.,* (2020). Submitted.
23. X.D. Niu, C. Shu, & Y.T. Chew, A thermal lattice Boltzmann model with diffuse scattering boundary condition for micro thermal flows. *Comput. Fluids,* 36, 273–281 (2007).
24. G.R. McNamara, & G. Zanetti, Use of the Boltzmann equation to simulate lattice-gas automata. *Phys. Rev. Lett.,* 61, 2332–2335 (1988).
25. N. Taoufik, Z. Jamil, & B.M. Rajeb, Lattice Boltzmann analysis of 2-d natural convection flow and heat transfer within square enclosure including an isothermal hot block. *Int J Thermal Tech.,* 3, 146–154 (2013).
26. N. Taoufik, & D. Ridha, Natural convection flow and heat transfer in square enclosure asymmetrically heated from below: A lattice Boltzmann comprehensive study. *Comput. Model. Eng. Sci.,* 88(3), 211–227 (2012).
27. L. Chen, Multi-scale modelling of proton exchange membrane fuel cell by coupling finite volume method and lattice Boltzmann method. *Int. J. Heat Mass Transf.,* 63, 268–283 (2013).
28. B. Han, & H. Meng, Numerical studies of interfacial phenomena in liquid water transport in polymer electrolyte membrane fuel cells using the lattice Boltzmann method. *Int. J. Hydrogen Energy,* 38, 5053–5059 (2013).
29. D. Froning, & J. Liang, Impact of compression on gas transport in non-woven gas diffusion layers of high temperature polymer electrolyte fuel cells. *J Power Sourc.,* 318, 26–34 (2016).
30. *ANSYS FLUENT, Theory Manual (Fuel Cells),* Version 19, Lebanon, NH, USA (2019).

9 Graphene-based Composites for High-Speed Energy Storage Battery Application

Satya Narayan Agarwal, Ashish Shrivastava, and Kulwant Singh

CONTENTS

9.1 Introduction 171
9.2 Basic Structure of Battery 173
9.3 Graphene Synthesis Techniques and Properties 173
9.4 Graphene-Based Anode 175
9.5 Graphene-Based Cathode 176
    9.5.1 Li–S Batteries 176
    9.5.2 Li–Air Batteries 177
    9.5.3 Others 177
9.6 Developing Properties of EES Devices Designed from Graphene-Based Composites 178
    9.6.1 Flexibility 178
    9.6.2 Transparency 178
    9.6.3 Free-Standing Property 179
    9.6.4 Fast-Charging Property for Batteries 179
    9.6.5 Other Properties 180
9.7 Outlook 180
References 181

9.1 INTRODUCTION

Increasing consumption of fossil fuel to fulfil current energy demand is a major concern for environmental safety. In the recently held COP26 summit, world leaders have decided to achieve net zero emission by mid-century. So, the problem of environment can be solved. Petrol and coal are the main fossil fuels which are used to fulfil power demand. Petroleum fuel demand, which is mainly used in the automobile sector, can be reduced by using electric vehicles. For electricity demand, solar can be the best solution but it is available in daytime for 6–12 hours depending on geographical location. This problem can be solved if large battery storage system can be installed. Some countries, such as Germany, Australia, and Switzerland, have already started working on building such types of energy storage systems. The solution for all the above global problems discussed is efficient energy storage design.

For efficient energy storage systems various types of batteries are available in market such as Li-ion, Na-ion, and Li-Cd batteries. Still the energy and power density of these batteries is limited, which needs improvement. To improve the energy and power density of available battery technology various efforts have been made by research community in the synthesis of advanced material for electrode which can improve the battery parameters. Graphene is one of the most important materials which can play a significant role in improvising energy and power density. Graphene possesses very good electrical properties such as charge mobility 20,000 cm^2/V-s, excellent thermal stability, conductivity (3,000–5,000 W/m-K), high Young's modulus (1 TPa), and large theoretical specific surface area (2,630 m^2/g) [1–3]. The addition of graphene in cell material improves the electrical parameters of cells [4].

All these properties make graphene a very good material to fabricate battery electrode for energy storage systems. To synthesize graphene and graphene-based derivate, the steps provided in Figure 9.1 are followed:

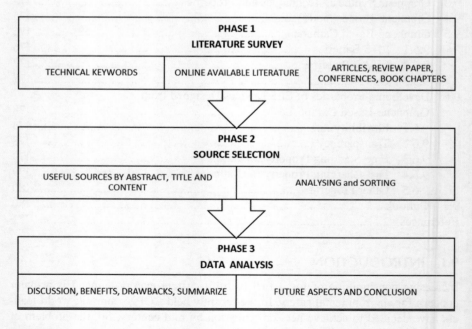

FIGURE 9.1 Methodology adopted for literature survey in this chapter.

9.2 BASIC STRUCTURE OF BATTERY

The basic structure of Li-ion batteries consists of anode, cathode, separator, electrolyte, and current collector (Figure 9.2). Separator is used to prevent touching of anode to cathode. Separator works as a filter which allows only Li-ion to propagate through while charging and discharging. Polyethylene or polypropylene is commonly used as the separator material. Anode of Li-ion battery cell is made of graphite which is used to store Li-ion while discharging and cathode is used to supply Li-ion. Cathode of Li-ion is made of $LiFePO_4$ and works as a positive electrode also.

The function of each part is described in Table 9.1.

9.3 GRAPHENE SYNTHESIS TECHNIQUES AND PROPERTIES

Graphene synthesis techniques are playing a very important role in application electrode fabrication for energy storage devices. There are various techniques available for graphene synthesis. However, they are classified into two main groups named as

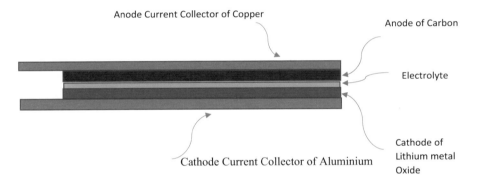

FIGURE 9.2 Basic structure of battery cell.

TABLE 9.1
Function of Each Part

Battery Part	Fabrication Material	Properties
Cathode	$LiFePO_4$	It works as positive electrode of a battery and provides lithium-ion which is responsible for charge storages.
Anode	Graphite	It works as negative electrode of a battery. It can store and release lithium-ion during charging and discharging process.
Electrolyte/separator	Polyethylene or polypropylene	It prevents contact between anode & cathode. It allows lithium-ion to pass. Separator is saturated with liquid electrolyte.
Current collector	Aluminium and copper	It provides support for anode and cathode.

top-down and bottom-up methods. In the top-down method mechanical exfoliation, arc discharge, oxidative exfoliation and reduction of GO (graphene oxide), liquid phase exfoliation, and unzipped of CNTs (carbon nanotubes) are the most widely used methods. Furthermore, oxidative exfoliation and reduction of GO method are divided into chemical reduction, thermal reduction, electrochemical reduction, and other reduction methods.

The first graphene synthesis was done in early 1975 by Lang et al. [5]. But, due to the discrepancy between synthesized graphene layered developed on various platinum crystal surfaces, it failed to identify the valuable applications of the product. After many efforts, graphene synthesis was successfully conducted in 1999 [6,7]. Although Novoselov et al. [8] was credited for the discovery of graphene in 2004, they first revealed the synthesis of graphene using exfoliation process. The method was followed with efforts to develop new techniques for effective and large-scale production of graphene. This method is known as mechanical exfoliation.

Graphene-based electrode designed from mechanically cleaved graphene, CVD-grown (chemical vapour deposition) graphene, or massively produced graphene derivatives from bulk graphite have been used in a variety of applications, including light-emitting diodes, touch screens, field-effect transistors, solar cells, supercapacitors, batteries, and sensors. They have been used as electrodes because of their low sheet resistance and high optical transmittance (Figure 9.3).

FESEM image of graphene oxide synthesis by modified hummer methods is shown in Figure 9.4. In Figure 9.4a–d, magnification factor (X) is 8,000, 9,000 and 7,000, but the layer of graphene oxide is not clear. In Figure 9.4b, at magnification factor (X) 35,000 graphene oxide layer has been observed.

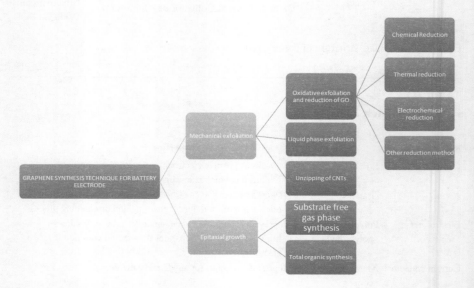

FIGURE 9.3 Method of graphene synthetic techniques for battery cell electrode.

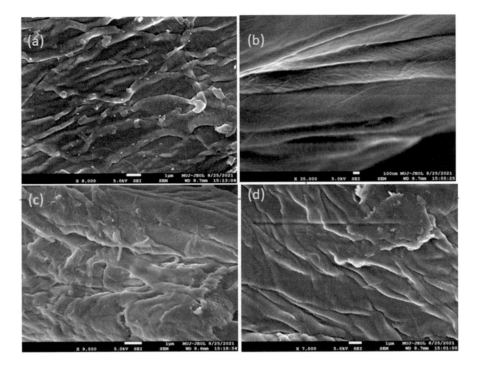

FIGURE 9.4 FESEM images of graphene oxide magnification synthesized using modified Hummer's method at different magnification factors displayed from (a) to (d).

9.4 GRAPHENE-BASED ANODE

Li-ion battery's theoretical capacity is 372 nmAh/g. Recently, various research articles have published showing increased energy and power density using graphene-based anode [9]. Graphene nanosheet obtained through an exfoliation achieved specific capacity of 540 mAh/g which can be further improved up to 730 mAh/g by incorporating CNT and C_{60} to graphene nanosheet. This improvement in energy density is due to the different electronic structure of graphene and conventional graphite used as anode in Li-ion batteries. Incorporation of graphene increases spacing between graphite and graphene nanosheet which allows it to accommodate more Li-ions. Other than graphene nanosheet, graphene paper can be used in flexible energy storage [10,11].

However, chemically prepared graphene shows high reversible capacity (1,264 mAh/g) than conventional graphite at low discharge rate.

Recently, heteroatom Nitrogen and Boron doped graphene has shown excellent energy density and power density. The author observed that improvement in performance using doped electrode is due to disordered surface morphology, better electrode/electrolyte wettability, improved electrical conductivity, and heteroatomic defects. However, successful synthesis of large-area monolayer and multilayer, and transferring it onto suitable substrate material is still a great challenge for the research community.

9.5 GRAPHENE-BASED CATHODE

In Li-ion batteries, graphene-based cathode enhances the performance. Graphene-based cathode shows high-energy, high-power density, long life, low discharge rate, etc. Graphene is defined as the one-atom-thick sheet of sp^2-bonded carbon atoms in a honeycomb crystal structure with a large theoretical specific area (2,600 m^2/g), proving its lithium storing capability. Graphene is known for its immense properties such as high surface area and short ion channel which results in the ability of Li-ion storage and transfers the conductivity and electric qualities which provide a solid basis for its electrochemical performance, outstanding stability, and expansion buffer, extending battery life. All of these benefits point to graphene's suitability as a fundamental material for electrodes and conductive additives in LIBs (lithium ion batteries). Recent research has indicated that graphene-based electrodes have greater specific capacities than many conventional electrode materials, demonstrating its potential as a helpful alternative. $LiFePO_4$/graphene composite, $LiMn_2O_4$/graphene composite, $Li_3V_2(PO_4)_3$/graphene composite can be used as the application for graphene as cathode materials in Li-ion batteries. Although there have been several discoveries of graphene-based cathodes in Li-ion batteries, active research should be focused on certain factors to meet the increasing need for a long lifespan, quick charge/discharge, and high specific capacity. Research should also be focused on practical use, rational design, and regulated cathode material preparation.

9.5.1 Li–S Batteries

Li–S batteries (LSBs) are particularly powerful in theoretical specific energy (2,600 Wh/kg or 2,800 Wh/L) by the use of electrochemical reversibility. LSBs have very high specific energy [12], a 3D-strength charge accumulator having submicron base and high active surface area [13], and high reversible electrochemical process using graphene oxide anode. In a recent research article, it has been observed that oxide exhibits the good lithiophilicity and active sites for the production of Li-ion. It also improves the charge accumulation on current collector [14,15]. The stability of foil is discussed in various atmospheres and display stable cyclability along with the protection of graphene sheets, i.e., 400 cycles with a capacity of 98% retention [16]. Furthermore, the group of Jiayan Luo developed an electric field detoured with the RGO (reduced graphene oxide)/Li composite anode and a horizontally centric lithium plate in the patterned spaces that could be cycled over 2,000 hours, with a stable voltage profile at a power density of 10 mA/cm^2 [17]. Sulphur cathode's advantages are abundance, cost-effective, environmental friendly and high specific capacity, i.e., 1,672 mAh/g. Its disadvantages are low electronic conductivity which leads to inactive electrochemical kinetics, the shuttle effect causes low-charge storage efficiency and poor life cycle, change of volume during charging and discharging process which gives rise to deterioration of the electrode [18]. To notice this problem, various graphene materials such as heteroatom-doped graphene [19], graphene oxide [20], reduced graphene oxide [21], porous graphene [22], graphene metal oxide [23], transition metal oxide, and graphene/micro/mesoporous carbon [24] have been used with various properties of chemical and physical for anode materials and the separators for soluble polysulphides to improve its shuttle effect [25–27]. Yuguo Guo's group [28] recently synthesized the composite S/(GCNs), which increased sulphur in it, which leads to improve its capacity, i.e., 765 mAh/g at 5°C, as well as its life cycle around thousand cycles due to strong conductivity and the physical constraints of the impact from the graphene–graphite carbon nanocage network. The composite produced showed strong stability of cycling, i.e., after hundred cycles, 85% capacity retention at 0.2°C and high initial capacity (1,471 mAh/g). Many graphene-based composites reported for LSBs which demonstrated an electrochemical performance in capacity that gained cycling performance compared with the Li-ion batteries, which is still very uncertain and further research needs to be improved for LSBs.

9.5.2 Li–Air Batteries

To fulfil the requirements of the rising energy demand, more efficient electrochemical energy storage (EES) devices which possess a higher energy density compared to LIBs must be developed. One of the most successful possibilities in this respect is Li–air batteries (LABs) with a hypothetical power density of 5,200 Wh/kg [29]. The oxygen evolution reaction process is the electrochemical decomposition of lithium peroxide or lithium oxide into lithium and oxygen [30,31]. The recharge of LABs has been researched in detail by Bruce and colleagues' pioneering experiments in 2006 [32].

Unfortunately, they are also seriously affected by their poor rate capability, short cycle lives, poor efficiencies, and production of irreversible lithium peroxide byproducts. Among the numerous aspects influencing the performance of LABs, air electrode (cathodes) has a substantial impact on the high energy density which is in high demand due to its catalytic activity with catalysts since it reduces its overall weight energy storage system for end-user products [29,31]. So, it seems to be an appropriate strategy to handle the issue for LABs using composites based on graphene catalyst with large specific surface area and porous nanostructures. Several transition metals/oxides such as platinum and nanostructured noble [33], Co_3O_4 [34], manganese cobalt oxide [35], and $Nd_{0.5}Sr_{0.5}CoO_{3-\delta}$ [36] have been merged with various materials of graphene and show better efficiency in LABs as compared with the simple-catalytic materials. The researchers have also shown that functional groups and imperfections in materials of graphene are shown as a catalyst in the development of disposal products [37,38].

9.5.3 Others

Recently, pristine graphene and its composites were reported using MgIBs (magnesium ion batteries) and AIBs (aluminium ion batteries) as electrodes, but electrochemical performance is still far short of realistic necessities [39,40]. Air electrodes for NABs [41,42], Zn–air batteries [43], vanadium redox-flow batteries (VRFBs) [44] were also reported for graphene-based composites to increase their performance. These inventions are relatively early for sodium air batteries, and zinc-air batteries require more development. In the case of VRFBs, the form of electrochemical energy storage is a promising device for practical applications, present procedures or techniques for electrodes and membranes synthetization are not able to meet high criteria. Graphene oxides and materials have been employed as additions to VRFB ion-exchange membranes to minimize vanadium permeability and prevent ion mixture [45]. However, for commercial applications, the techniques for the use of VRFB graphene-based composites are also too late.

9.6 DEVELOPING PROPERTIES OF EES DEVICES DESIGNED FROM GRAPHENE-BASED COMPOSITES

9.6.1 FLEXIBILITY

The demand for flexible electronics is quickly increasing in the field of solar cells to displays and wearable electronics [46]. The growth of this industry is largely attributable to more customer demand and rapid advancement in flexible materials research [46]. All capacitors and commercial batteries are unbendable and inflexible in the electro-chemical energy storage because of bending them to any large range might create electrolyte leakage or structural damage. This benefit might further be derived from graphene, a material that is believed to be an excellent flexible material because of its intrinsic features. Many flexible EES systems are based on these materials like twistable [47], wearable [48], ECs (electro-chemical capacitors) and LIBs rollable [49], and stretchable [50] with adjustable shapes, dimensions, and its properties are being studied [51–53]. The wearable electrochemical energy storage devices, which may be integrated into clothes and accessories for wearable electronics, are quite desirable amongst the several flexible EES systems [54]. These intelligent clothing are expected to provide consumers with real-time information [55]. The group of Zhenan Bao initially created ECs based on textile in 2011 by using graphene/MnO_2 on porous textile fibres [56]. In recent years, good research have been focused on the creation of core sheaths and coaxial electrodes that combine the current collector and active materials in the same thread that may be knitted or woven into textiles directly [57]. For example, Chao Gao's group reported high-flexibility, high-weavability composite yarns made of polyelectrolyte-wrapped graphene, which might have a significant improvement in the area capacity to 269 mF/cm^2 [58].

9.6.2 TRANSPARENCY

There has been considerable interest in transparent devices. Different transparent devices have been shown in many industries, including smart windows, solar cells, touchscreens, and so on using the advances of transparent materials and technology research. Nevertheless, the progress of transparent power like a fundamental constituent for integrated transparent devices remains a problem, likely as the transparency of each component is difficult to achieve simultaneously [59]. Many classic layer structure for power source systems (e.g., LIB) does include cathode and anode electrodes, electrolyte, separator, packing, etc. One method is to decrease its thickness into the nanoscale scale to build a transparent electrode; however, this plan would compromise most of its capacity [60]. The second method is to construct a designed electrode with its characteristic size below the human eye's resolution limit [61].

Zarbin's group employed a unique single-pot synthesis to develop composite films made of graphene/polyaniline that was very transparent and flexible and could readily be transferred for numerous common substrate applications [62,63]. When this composite is used to make a transparent supercapacitor device, this composite may exhibit good pseudocapacitive with a specific capacitance of 267.2 F/cm^3. Graphene-based composites are likely to have an increasingly major role in the future in the area of transparent power source systems.

9.6.3 Free-Standing Property

In general, it is the combination of particulate polymeric binders, e.g., polytetrafluoroethylene and active agents (e.g., carbon black) in appropriate solvents that is used to fabricate commercial EES electrodes like capacitors and batteries. The slurries formed are covered with suitable current collectors. Usually, Cu foils are used for the anode, and aluminium foils are used for the cathode. The electrodes are then dried and pressed mechanically to modify the volume and density of the pore. For free-standing, electrodes such as graphene/silicon film [64] and graphene/metal oxide [65] are the graphene metal composites used for energy storage devices. Likewise, a cathode made up of graphene metal composites used in LFP (lithium iron phosphate batteries) shows strong mechanical flexibility that could generate with a specific capacity of approx. 80 mAh/g at 100°C, despite the relatively high mass loading condition (6 mg/cm^2) [66]. Future free-standing electrochemical energy storage appliances based on graphene-based composites are predicted to be more operative with better energy density.

9.6.4 Fast-Charging Property for Batteries

Battery systems such as LIBs with the slow charge transfer rate which concluded the long charging time in few minutes to hours while the customer demands the shorter charging duration. In 1991, the first commercialization was developed by the Sony organization, where they took much time to recharge its battery status [67]. Researchers demonstrated the 3D porous N-doped graphene aerogel or lithium iron phosphate cathode composite having the charge capacity of 145 mAh/g in 60 seconds with a charge efficiency of 97% [66]. Hui-Ming Cheng's group demonstrated the titanium dioxide quantum dot graphene nanosheet anode composite which resulted in the success of charge and discharge at 50°C which is equivalent to fast charging and discharging process in 72 seconds [68]. Moreover, the Hui-Ming Cheng group reported the LIB fabricated by using the graphene foam or lithium titanate anode material and graphene foam or lithium iron phosphate cathode material without the use of binders or conductive agents [69]. The proposed Li-ion batteries could be fully charged in about 18 seconds because of their open porous structure and electronic conductivity. Other graphene-based composites battery solutions have demonstrated the ability to charge quickly. For example, to achieve the fast charging and discharging process at 10°C, the Qiang Zhang group reported the unstacked double-layer template for cathode material, i.e., graphene/sulphur composite for LSBs [70].

Basically, the rapid-charging feature of graphene-based composites is not restricted to batteries; it may also be used to increase the quick charge (and discharge) properties of capacitors. However, micro-supercapacitors deliver the higher volumetric energy density in ionic liquid about 2.42 mWh/cm^3, which is in more than two orders of magnitude than AECs, which makes the possibility to make the lesser and low-weight electronic devices [71–73].

9.6.5 OTHER PROPERTIES

Besides the above section, from graphene-based composites other unique properties for the EES appliances were discovered like solid-state [74], self-healing [75,76], and all graphene-based structure [77–79]. Zhanhu Guo's group discovered the electrochromic polyaniline/GO nanocomposite film with durable electrochemical energy storage performance [80]. Coloration efficiency and an areal capacitance are 59.3 cm^2/C^1 and 25.7 mF/cm^2, respectively. With a 100% stretching capacity the supercapacitor has 82.4% capacitance retention and 52.4% have the third healing capacitance rate [81–83]. Furthermore, Wencai Ren's group fabricated the integrated micro-supercapacitors which are based on various substrates of in-series screen-printed graphene, exceptional electrical double-layer capacitive behaviours with metal-free current collectors, high-voltage output, and outstanding flexibility [84]. Hui-Ming Cheng's fabrication for fast charging and high-energy density for LSB cathode, some sandwiched like graphene-based structure, has been generated [85,86]. Additionally, Chunsheng Wang's group [87] and Xinliang Feng et al. [4] designed all-solid-state LSBs and pseudo-capacitors which depends on graphene materials having superior electrochemical performance and safety.

9.7 OUTLOOK

In this chapter, the advancement in graphene-based energy storage devices has been discussed based on interfacial engineering and mechanism for various applications. Graphene-as composites material for EES has shown significant advancements and quick growth nowadays and numerous industry level problems can be solved. Firstly, the absence of realistic mass manufacturing procedures of synthesis for high-quality graphene and its metal composites used in electrochemical energy storage sector. As a result, the synthesis of graphene and graphene-based composite will lead to the preparation for innovative material which proceeds with the appropriate characteristics and will promote to attract a lot of attention in the future. Secondly, the large surface area of graphene-based composites is important for their use in supercapacitors, including EDLCs (electro-chemical double layer capacitors) and pseudo-capacitors. These composites have increased supercapacitor capacities due to their high surface area, higher electrical conductivity, and suitable porous architectures. However, the porous structure reduces its volumetric energy density dramatically. By improving their porous structure in order to commercialize them in supercapacitors, the packing densities can be improved. Thirdly, graphene-based composites had been extensively examined in different metal-ion and metal-sulphur batteries like LIBs and NIBs (sodium ion batteries) with outstanding electrochemical performance, their high electrical conductivity, surface area, special nanoporous architecture, and so on.

However, in most situations, the graphene component's mass ratio is relatively high in these composites, often greater than 10 wt%, lowering the energy density and packing density of the batteries. As a result, greater effort should be put into optimization between charge transport and their rapid diffusion. Fourth, metal-air batteries can do revolution in energy storage market as LABs have shown extraordinarily high theoretical energy density. While numerous graphene-based composites have been synthesized with increased performance (such as graphene/platinum and graphene/ruthenium), the obtained capabilities are not near the theoretical energy density. To achieve the purpose in the practical application of graphene-based composites, it becomes important that it should be focused on in future work. Hence, to enhance the research achievement in the industry, the electrochemical performance is shown which would be considered for graphene-based composites.

REFERENCES

1. M. Armand, J.M. Tarascon, *Nature* 451 (2008) 652–657.
2. J.M. Tarascon, M. Armand, *Nature* 414 (2001) 359–367.
3. J. Meng, H. Guo, C. Niu, Y. Zhao, L. Xu, Q. Li, L. Mai, *Joule*, 1, 522–547.
4. Z.-S. Wu, Y. Zheng, S. Zheng, S. Wang, C. Sun, K. Parvez, T. Ikeda, X. Bao, K. Müllen, X. Feng, *Adv. Mater.* 29 (2017) 1602960.5. B.A. Lang, *Surf. Sci.* 53(1) (1975) 317e29.
5. E. Rokuta, Y. Hasegawa, A. Itoh, K. Yamashita, T. Tanaka, S. Otani, et al. *Surf. Sci.* 427 (1999) 97e101.
6. H. Shioyama, *J. Mater. Sci. Lett.* 20(6) (2001) 499e500.
7. K.S. Novoselov, D Jiang, F. Schedin, T. Booth, V. Khotkevich, S. Morozov, et al. *Proc. Natl. Acad. Sci. U. S. A.* 102(30) 2005 10451e3.
8. C. Xu, B. Xu, Y. Gu, Z. Xiong, J. Sun, X.S. Zhao, *Energy. Environ. Sci.* 6 (2013) 1388–1414.
9. S. Chowdhury, R. Balasubramanian, *Prog. Mater. Sci.* 90 (2017) 224–275.
10. J. Zang, S. Ryu, N. Pugno, Q. Wang, Q. Tu, M.J. Buehler, X. Zhao, *Nat. Mater.* 12 (2013) 321.
11. K.S. Novoselov, A.K. Geim, S.V. Morozov, D. Jiang, M.I. Katsnelson, I.V. Grigorieva, S.V. Dubonos, A.A. Firsov, *Nature* 438 (2005) 197.
12. M. Terrones, A.R. Botello-Mendez, J. Campos-Delgado, F. Lopez-Urias, Y.I. Vega-Cantu, F.J. Rodríguez-Macías, A.L. Elías, E. Munoz-Sandoval, A.G. Cano-Marquez, J.-C. Charlier, H. Terrones, *Nano Today* 5 (2010) 351–372.
13. Y. Zhang, Y.-W. Tan, H.L. Stormer, P. Kim, *Nature* 438 (2005) 201.
14. J.C. Meyer, A.K. Geim, M.I. Katsnelson, K.S. Novoselov, T.J. Booth, S. Roth, *Nature* 446 (2007) 60.
15. J. Zhao, G. Zhou, K. Yan, J. Xie, Y. Li, L. Liao, Y. Jin, K. Liu, P.-C. Hsu, J. Wang, H.-M. Cheng, Y. Cui, *Nat. Nanotechnol.* 12 (2017) 993.
16. A. Wang, X. Zhang, Y.-W. Yang, J. Huang, X. Liu, J. Luo, *Chem* 4 (2018) 2192–2200.
17. Q. Pang, X. Liang, C.Y. Kwok, L.F. Nazar, *Nat. Energy* 1 (2016) 16132.
18. J. Song, Z. Yu, M.L. Gordin, D. Wang, *Nano Lett.* 16 (2016) 864–870.
19. Z. Yunbo, M. Lixiao, N. Jing, X. Zhichang, H. Long, W. Bin, Z. Linjie, *2D Mater.* 2 (2015) 024013.
20. N. Li, M. Zheng, H. Lu, Z. Hu, C. Shen, X. Chang, G. Ji, J. Cao, Y. Shi, *Chem. Commun.* 48 (2012) 4106–4108.
21. X. Yang, L. Zhang, F. Zhang, Y. Huang, Y. Chen, *ACS Nano* 8 (2014) 5208–5215.
22. Z. Xiao, Z. Yang, L. Wang, H. Nie, M.e. Zhong, Q. Lai, X. Xu, L. Zhang, S. Huang, *Adv. Mater.* 27 (2015) 2891–2898.

23. H.-J. Peng, J.-Q. Huang, M.-Q. Zhao, Q. Zhang, X.-B. Cheng, X.-Y. Liu, W.-Z. Qian, F. Wei, *Adv. Funct. Mater.* 24 (2014) 2772–2781.
24. G. Liu, W. Jin, N. Xu, *Chem. Soc. Rev.* 44 (2015) 5016–5030.
25. Y. Yuan, B. Wang, R. Song, F. Wang, H. Luo, T. Gao, D. Wang, *New J. Chem.* 42 (2018) 6626–6630.
26. R.-S. Song, B. Wang, T.-T. Ruan, L. Wang, H. Luo, F. Wang, T.-T. Gao, D.-L. Wang, *Appl. Surf. Sci.* 427 (2018) 396–404. Part B.
27. J. Zhang, C.-P. Yang, Y.-X. Yin, L.-J. Wan, Y.-G. Guo, *Adv. Mater.* 28 (2016) 9539–9544.
28. P.G. Bruce, S.A. Freunberger, L.J. Hardwick, J.-M. Tarascon, *Nat. Mater.* 11 (2012) 19–29.
29. H. Liu, M. Jia, Q. Zhu, B. Cao, R. Chen, Y. Wang, F. Wu, B. Xu, *ACS Appl. Mater. Interf.* 8 (2016) 26878–26885.
30. G. Girishkumar, B. McCloskey, A.C. Luntz, S. Swanson, W. Wilcke, *J. Phys. Chem. Lett.* 1 (2010) 2193–2203.
31. T. Ogasawara, A. Debart, M. Holzapfel, P. Novak, P.G. Bruce, *J. Am. Chem. Soc.* 128 (2006) 1390–1393.
32. Y. Yang, M. Shi, Q.-F. Zhou, Y.-S. Li, Z.-W. Fu, *Electrochem. Commun.* 20 (2012) 11–14.
33. W.-H. Ryu, T.-H. Yoon, S.H. Song, S. Jeon, Y.-J. Park, I.-D. Kim, *Nano Lett.* 13 (2013) 4190–4197.
34. H. Wang, Y. Yang, Y. Liang, G. Zheng, Y. Li, Y. Cui, H. Dai, *Energy Environ. Sci.* 5 (2012) 7931–7935.
35. C. Kim, O. Gwon, I.-Y. Jeon, Y. Kim, J. Shin, Y.-W. Ju, J.-B. Baek, G. Kim, *J. Mater. Chem. A* 4 (2016) 2122–2127.
36. J. Xiao, D. Mei, X. Li, W. Xu, D. Wang, G.L. Graff, W.D. Bennett, Z. Nie, L.V. Saraf, I.A. Aksay, J. Liu, J.-G. Zhang, *Nano Lett.* 11 (2011) 5071–5078.
37. B. Sun, B. Wang, D. Su, L. Xiao, H. Ahn, G. Wang, *Carbon* 50 (2012) 727–733.
38. Q. An, Y. Li, H. Deog Yoo, S. Chen, Q. Ru, L. Mai, Y. Yao, *Nano Energy* 18 (2015) 265–272.
39. H. Chen, H. Xu, S. Wang, T. Huang, J. Xi, S. Cai, F. Guo, Z. Xu, W. Gao, C. Gao, *Sci. Adv.* 3 (2017) eaao7233.
40. H. Huang, F. Zhou, X. Shi, J. Qin, Z. Zhang, X. Bao, Z.-S. Wu, *Energy Storage Mater.* (2019). Doi: 10.1016/j.ensm.2019.03.001.
41. Z. Khan, S. Park, S.M. Hwang, J. Yang, Y. Lee, H.-K. Song, Y. Kim, H. Ko, *NPG Asia Mater.* 8 (2016) e294.
42. M. Zeng, Y. Liu, F. Zhao, K. Nie, N. Han, X. Wang, W. Huang, X. Song, J. Zhong, Y. Li, *Adv. Funct. Mater.* 26 (2016) 4397–4404.
43. P. Han, Y. Yue, Z. Liu, W. Xu, L. Zhang, H. Xu, S. Dong, G. Cui, *Energy Environ. Sci.* 4 (2011) 4710–4717.
44. M.A. Aziz, S. Shanmugam, *J. Mater. Chem. A* 6 (2018) 17740–17750.
45. A. Nathan, A. Ahnood, M.T. Cole, S. Lee, Y. Suzuki, P. Hiralal, F. Bonaccorso, T. Hasan, L. Garcia-Gancedo, A. Dyadyusha, S. Haque, P. Andrew, S. Hofmann, J. Moultrie, D. Chu, A.J. Flewitt, A.C. Ferrari, M.J. Kelly, J. Robertson, G.A.J. Amaratunga, W.I. Milne, *Proc. IEEE* 100 (2012) 1486–1517.
46. Z. Li, X. Liu, L. Wang, F. Bu, J. Wei, D. Pan, M. Wu, *Small* 14 (2018) 1801498.
47. L. Liu, Y. Yu, C. Yan, K. Li, Z. Zheng, *Nat. Commun.* 6 (2015) 7260.
48. J. Yang, J. Zhang, X. Li, J. Zhou, Y. Li, Z. Wang, J. Cheng, Q. Guan, B. Wang, *Nano Energy* 53 (2018) 916–925.
49. B. Qiu, M. Xing, J. Zhang, *J. Mater. Chem. A* 3 (2015) 12820–12827.
50. Y. Wang, X. Yang, L. Qiu, D. Li, *Energy Environ. Sci.* 6 (2013) 477–481.
51. H.-P. Cong, X.-C. Ren, P. Wang, S.-H. Yu, *Energy Environ. Sci.* 6 (2013) 1185–1191.
52. Y. Meng, K. Wang, Y. Zhang, Z. Wei, *Adv. Mater.* 25 (2013) 6985–6990.

Graphene-based Composites   183

53. Y.-H. Lee, J.-S. Kim, J. Noh, I. Lee, H.J. Kim, S. Choi, J. Seo, S. Jeon, T.-S. Kim, J.-Y. Lee, J.W. Choi, *Nano Lett.* 13 (2013) 5753–5761.
54. Y. Zheng, X. Ding, C.C.Y. Poon, B.P.L. Lo, H. Zhang, X. Zhou, G. Yang, N. Zhao, Y. Zhang, *IEEE (Inst. Electr. Electron. Eng.) Trans. Biomed. Eng.* 61 (2014) 1538–1554.
55. G. Yu, L. Hu, M. Vosgueritchian, H. Wang, X. Xie, J.R. McDonough, X. Cui, Y. Cui, Z. Bao, *Nano Lett.* 11 (2011) 2905–2911.
56. S.H. Aboutalebi, R. Jalili, D. Esrafilzadeh, M. Salari, Z. Gholamvand, S. Aminorroaya Yamini, K. Konstantinov, R.L. Shepherd, J. Chen, S.E. Moulton, P.C. Innis, A.I. Minett, J.M. Razal, G.G. Wallace, *ACS Nano* 8 (2014) 2456–2466.
57. L. Kou, T. Huang, B. Zheng, Y. Han, X. Zhao, K. Gopalsamy, H. Sun, C. Gao, *Nat. Commun.* 5 (2014) 3754.
58. J. Ge, G. Cheng, L. Chen, *Nanoscale* 3 (2011) 3084–3088.
59. T. Chen, Y. Xue, A.K. Roy, L. Dai, *ACS Nano* 8 (2014) 1039–1046.
60. Y. Yang, S. Jeong, L. Hu, H. Wu, S.W. Lee, Y. Cui, *Proc. Natl. Acad. Sci. U.S.A.* 108 (2011) 13013–13018.
61. V.H.R. Souza, M.M. Oliveira, A.J.G. Zarbin, *J. Power Sources* 348 (2017) 87–93.
62. T.-K. Hong, D.W. Lee, H.J. Choi, H.S. Shin, B.-S. Kim, *ACS Nano* 4 (2010) 3861–3868.
63. J.-Z. Wang, C. Zhong, S.-L. Chou, H.-K. Liu, *Electrochem. Commun.* 12 (2010) 1467–1470.
64. L. Kong, C. Zhang, J. Wang, W. Qiao, L. Ling, D. Long, *ACS Nano* 9 (2015) 11200–11208.
65. B. Wang, Y. Xie, T. Liu, H. Luo, B. Wang, C. Wang, L. Wang, D. Wang, S. Dou, Y. Zhou, *Nano Energy* 42 (2017) 363–372.
66. Y. Nishi, in: G. Pistoia (Ed.), *Lithium-Ion Batteries*, Elsevier, Amsterdam, 2014, 21–39.
67. R. Mo, Z. Lei, K. Sun, D. Rooney, *Adv. Mater.* 26 (2014) 2084–2088.
68. N. Li, Z. Chen, W. Ren, F. Li, H.-M. Cheng, *Proc. Natl. Acad. Sci. U.S.A.* 109 (2012) 17360–17365.
69. M.-Q. Zhao, Q. Zhang, J.-Q. Huang, G.-L. Tian, J.-Q. Nie, H.-J. Peng, F. Wei, *Nat. Commun.* 5 (2014) 3410.
70. J. Yan, Q. Wang, T. Wei, Z. Fan, *Adv. Energy Mater.* 4 (2014) 1300816.
71. T.-W. Lin, C.-S. Dai, K.-C. Hung, *Sci. Rep.* 4 (2014) 7274.
72. M.F. El-Kady, M. Ihns, M. Li, J.Y. Hwang, M.F. Mousavi, L. Chaney, A.T. Lech, R.B. Kaner, *Proc. Natl. Acad. Sci. U.S.A.* 112 (2015) 4233–4238.
73. H. Gao, F. Xiao, C.B. Ching, H. Duan, *ACS Appl. Mater. Interf.* 4 (2012) 7020–7026.
74. K. Qin, L. Wang, S. Wen, L. Diao, P. Liu, J. Li, L. Ma, C. Shi, C. Zhong, W. Hu, E. Liu, N. Zhao, *J. Mater. Chem. A* 6 (2018) 8109–8119.
75. Y. Yue, N. Liu, Y. Ma, S. Wang, W. Liu, C. Luo, H. Zhang, F. Cheng, J. Rao, X. Hu, J. Su, Y. Gao, *ACS Nano* 12 (2018) 4224–4232.
76. N. Li, Z. Chen, W. Ren, F. Li, H.-M. Cheng, *Proc. Natl. Acad. Sci. U.S.A.* 109 (2012) 17360–17365.
77. G. Zhou, S. Pei, L. Li, D.-W. Wang, S. Wang, K. Huang, L.-C. Yin, F. Li, H.-M. Cheng, *Adv. Mater.* 26 (2014) 625–631.
78. R. Fang, S. Zhao, S. Pei, X. Qian, P.-X. Hou, H.-M. Cheng, C. Liu, F. Li, *ACS Nano* 10 (2016) 8676–8682.
79. H. Wei, J. Zhu, S. Wu, S. Wei, Z. Guo, *Polymer* 54 (2013) 1820–1831.
80. C. Teng, D. Xie, J. Wang, Z. Yang, G. Ren, Y. Zhu, *Adv. Funct. Mater.* (2017) 1700240.
81. J.-Q. Huang, T.-Z. Zhuang, Q. Zhang, H.-J. Peng, C.-M. Chen, F. Wei, *ACS Nano* 9 (2015) 3002–3011.
82. H. Gao, F. Xiao, C.B. Ching, H. Duan, *ACS Appl. Mater. Interf.* 4 (2012) 7020–7026.
83. G. Zhou, S. Pei, L. Li, D.-W. Wang, S. Wang, K. Huang, L.-C. Yin, F. Li, H.-M. Cheng, *Adv. Mater.* 26 (2014) 625–631.
84. R. Fang, S. Zhao, S. Pei, X. Qian, P.-X. Hou, H.-M. Cheng, C. Liu, F. Li, *ACS Nano* 10 (2016) 8676–8682.

86. X. Shi, S. Pei, F. Zhou, W. Ren, H.-M. Cheng, Z.-S. Wu, X. Bao, *Energy Environ. Sci.* 12 (2019) 1534–1541.
87. X. Yao, N. Huang, F. Han, Q. Zhang, H. Wan, J.P. Mwizerwa, C. Wang, X. Xu, *Adv. Energy Mater.* 7 (2017) 1602923.

10 Energy Applications of Ionic Liquids

*Moumita Saha, Manoj K. Banjare,
Kamal Nayan Sharma, Gyandshwar K. Rao,
Anirban Das, Monika Vats, Gaurav Choudhary,
Kamalakanta Behera, and Shruti Trivedi*

CONTENTS

10.1 Introduction	185
10.2 Li-Ion Batteries	186
10.3 Fuel Cells	189
10.4 Solar Cells	196
10.5 Conclusions	200
References	201

10.1 INTRODUCTION

Currently, the energy demand worldwide is very high. In order to fulfil it we are more dependent on the fossil fuels, but limitation of fossil fuels, non-renewability, and the environmental hazards are a big concern. So, it demands our attention towards it, and upgradation towards renewable energy use is required. Hence there is an urgent demand for the development of alternative novel approaches to energy generation and storage with long-term sustainability. As renewable energies are sporadic, they required good production, storage, and delivery systems. Renewable energy (like solar, wind, wave, etc.) is almost available everywhere; so, it can potentially replace non-renewable energies and compensate heavy energy demand with a more environmental-friendly approach. As in the current scenario, increasingly greater wind and solar energy capacity are installed worldwide and it requires very large-scale solutions for energy storage. Further, researchers across the globe are putting their best efforts to minimize the impact of largely used fossil fuels through various technologies for carbon capture. Currently, a vast amount of effort is made towards the discovery of novel materials that can show immense potential towards energy generation, storage, and delivery. In green chemistry, ionic liquids (ILs) open a new path. ILs have shown great impact on various energy technologies, and more importantly, offer a wide range of physicochemical properties which can be effectively tuned to obtain improved performance in those energy technologies [1–13].

ILs are a new class of compounds which are like salts but generally liquid in room temperature (below 100°C). Just as its name suggests ILs are completely ionic. Here, in ILs one of the ions (generally cations) is organic moiety. For example, ILs are composed of bulky cations like alkylammonium, alkylphosphonium, N-alkyl imidazolium, alkylpyridinium, pyrrolidinium, and sulfonium with anions like $[BF_4]^-$, $[PF_6]^-$, $[CF_3SO_3]^-$, and $[(CF_3SO_2)_2N]^-$. ILs have some unique properties that attract the attention of many researchers from interdisciplinary research areas towards it. More importantly their properties are tunable, which means anyone can design their own ILs in their desired way. Minute changes in the structure of the cation and/or in anion changes the physicochemical properties of the ILs. Some common properties shown by ILs are high viscosity, low melting point, and high boiling point, almost negligible vapour pressure, high thermal and electrical stability, and broad electrochemical window. ILs are reported to be used in almost every sector, especially as solvents (even reusable after proper purification with no noticeable difference in quality) in no matter organic or inorganic reactions, as catalyst, electrolyte in energy storage devices as well as in conversion devices. This chapter highlights the use of ILs in various energy converting devices and energy storing devices, such as fuel cells, solar cells, and batteries [1–131].

10.2 Li-ION BATTERIES

ILs have shown tremendous application potential as electrolytes in Li-ion batteries (LIBs). LIBs mainly consist of three major components, such as cathode, anode, and electrolyte, which are a subject of extensive research. Generally, metal oxide or specifically lithium compounds are used for cathode and carbon, and recently graphite is used for anode. But recently a lot of metal oxides and other compounds are explored for cathode ($NaFeO_2$, spinel and olivine structures, $LiFePO_4$, $LiCoO_2$) [1] and transition metal oxide, silicon as anode ($ZnCo_2O_4$, $MnCo_2O_4$, $CoMn_2O_4$, $NiCo_2O_4$, Fe_3O_4, Co_3O_4, Fe_2O_3, NiO, CoO, MnO, Mn_2O_3, Cr_2O_3, Mn_3O_4, MoO_3, MnO_2) [2]. Better performance of LIBs was found with nano $Li_3V_2(PO_4)_3$ [3]. Commercially mixture of organic carbonate like ethylene carbonate, diethyl carbonate, vinylene carbonate and lithium complex like $LiPF_6$, $LiAsF_6$, $LiClO_4$, $LiBF_4$, $LiCF_3SO_3$ are used [4] (Figure 10.1).

At the discharge time Li ion moves from anode to cathode through the electrolyte and opposite for charging. LIBs have low self-discharge and high energy density. LIBs provide the advantages like longer life, better capacity, temperature tolerant, fast charging, maintenance free and light weight. But it still needs the improvement in safety, cost, charging speed, working temperature range, etc. Organic electrolytes are good ionic conductors but flammable and volatile which reduces the operative temperature, with which $LiPF_6$ (generally) is added because it has ability to form layer on aluminium current collector. Organic electrolyte with $LiPF_6$ degrade over the voltage 4.3 V [6]. Sometimes additives (vinylene carbonate [7]) are also added in trace amounts (<5%). Important features that need to be considered while choosing an electrolyte for LIBs are favourable transport properties, chemical and

Energy Applications of Ionic Liquids 187

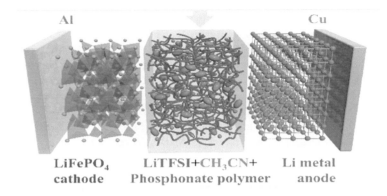

LiFePO$_4$ cathode   LiTFSI+CH$_3$CN+ Phosphonate polymer   Li metal anode

FIGURE 10.1 Schematic diagram of an LIB based on polymer electrolytes [5]. (Reproduced with permission from Brassinne et al. (Ref. 5). Copyright 2021 American Chemical Society.)

electrochemical stability, low melting point and high boiling point, and low vapour pressure [8]. The cathode and anode of LIBs are not stable with respect to electrolytes and in order to insert and extract Li ion between graphite layer a passive layer is important. Formation of a protective layer on the surface of electrode by electrolyte is called solid electrolyte interface (SEI). LIBs are no exception. On the surface of anode of graphite, a Li conducting film is formed which prevents the anode from further degradation. Formation of this SEI is very much important in the stability and longevity of a battery because without SEI deterioration of anode and battery dies [9]. So, whenever any new electrode or electrolyte is tested for LIBs, SEI is taken into consideration.

While considering an IL as electrolyte one has to contemplate its physical properties like viscosity, conductivity, and thermal stability. The combination of cation and anion can differ in these properties. Imidazolium, tetraalkylammonium and cyclic aliphatic quaternary ammonium were found to be the most viable cations. DMIM shows higher viscosity than MIM due to its larger size. Compared with imidazolium and ammonium ILs, pyrrolidium and pipyridinium ILs exhibit higher ESW because of nonaromatic character and are widely used. Introduction of electron donating group in side chain increases its voltage stability. But extreme increase in chain has no impact due to the little influence on electron donating ability [10]. Even incorporation of heteroatoms such as 'O' or 'S' in cation alkyl side chain was seen to affect the physicochemical property [11,12]. ILs with small anion can produce high I_c [13]. Among the anion families the perfluoroalkylsulfonylimmide is the most promising class of ILs [14]. TFSI and FSI are also good enough to mention. Also, FSI shows good film forming ability (SEI) [15,16]. Thus, EMIFSI has low viscosity and high conductivity than EMITFSI. But thermal stability of EMITFSI is greater than EMIFSI [17]. But neat ILs have high viscosity, which is not suitable for LIBs, so another solvent is generally used to dilute and reduce the viscosity. The shared solvent can be any molecular organic solvent, another IL, even water [18]. Brutti et al. compared EMIFSI, EMITFSI, N_{1114}FSI, N_{1114}TFSI, $N_{111(2O1)}$ TFSI, $N_{122(2O1)}$ TFSI. EMI-based ILs exhibit better conductivity than N_{1114}-based ILs. Addition of ether 'O' in the alkyl side chain affect the transport number depending on the cation structure and nature. But N_{1114} ILs were more stable than EMI. When LIB was prepared both 0.2 LiTFSi+0.8 IL composition delivers capacity of ~200 mA h/g. But EMI-based ILs are able to support a prolonged and stable cycle where N_{1114} degrade [19], whereas conductivity of DEDMIBF$_4$ is lower than DEDMITFSI and also melting point of DEDMITFSI is almost 20°C lower than DEDMIBF$_4$. When IL DEDMITFSI was used to prepare two electrolytes 0.4 LiTFSI/DEDMITFSI and 0.8 LiTFSI/DEDMITFSI, the order of specific conductivities was found to be DEDMITFSI > 0.4 M LiTFSI/DEDMITFSI > 0.8 M LiTFSI/DEDMITFSI [20]. Among ILs, aprotic ILs are majorly reported to be used as an electrolyte in LIBs. APILs show large ESW, high thermal stability, and low vapour pressure. When PYR$_{13}$TFSI is mixed with LiTFSI and organic additives, it shows great performance with LiFePO$_4$/Li cells. With different concentrations of LiTFSI, the prepared cell shows good reversible capacity at room temperature in the range 152–157 mA h/g at 0.1° C. Increasing concentration of LiTFSI was found to have an impact on the coulombic efficiency of the cell too [21].

Not only aprotic ILs but also protic ILs can be used as electrolytes for LIBs. The main obstacle with APILs is their cost but this problem is somewhat solved by PILs, as PILs are easy to synthesize and cheap. When PILs are compared with APILs, the thermal stability of PILs is lower because of the protic hydrogen but that is obviously higher than organic electrolytes. Both PILs and APILs show comparable conductivity but APILs are more viscous than PILs, but PILs show smaller ESW than APILs [22]. Menne et al. for the first time reported the use of PILs as the electrolyte in LIBs. They used 1 M LiTFSI in triethylammonium bis(tetrafluoromethylsulfonyl)amide (Et$_3$NHTFSI) with LTO as anode and LFP as cathode. The electrolyte shows comparable conductivity with APILs and high ESW. The LIB shows 115 mA h/g and 30 mA h/g capacity for 0.1° C and 10° C, respectively. At room temperature it delivers 60 mA h/g specific capacitance for 1° C and stable up to 300 cycles [23]. When DMIM and MIM with FSI and TFSI were compared, MIMFSI shows the highest conductivity whereas MIMTFSI shows the lowest. Among them at 0.5° C LFP electrode displays specific capacitance 158 mA h/g for 0.5 M LiTFSI/DMIMTFSI. It retains more than 90% of initial capacitance at 1° C [24]. Eutectic mixture of PILs can also be used as electrolytes. Like 7:3 mixture of PYR$_{H4}$TFSI (mp 30° C) and PYR$_{HH}$TFSI (mp 37° C) reduces the melting point to –81° C. At 40° C the trio combination 0.5 LiTESI in PYR$_{HH}$TFSI and PYR$_{H4}$TFSI (7:3) show specific discharge capacity in LFP electrode more than 140 mA h/g at current density to 1° C, and 60 mA h/g at 10° C [25].

A new solid-state LIB was prepared using LiSICON as electrolyte, graphite as anode, and LiCoO$_2$ as cathode. LiSICON was obtained from ceramic. LiTFSI dissolved in PYR$_{13}$FSI was used as a wetting agent. The cell was tested with different LiTFSI concentration. A 'C' anode in a LiSICON cell with 0.75 M LiTFSI/PYR$_{13}$FSI had capacity of 325 mA h/g and coulombic efficiency of 99.8% during cycling at 80°C [26]. Redox elements are also reported to be used in corporation with IL electrolytes [27]. Balducci et al. reported comparison of PYR$_{14}$TFSI and PYR$_{14}$FSI-based liquid

(0.9PYR$_{14}$TFSI–0.1LiFSI) and polymer electrolyte (cl-PEO–PYR$_{14}$TFSI–LiTFSI, PIL–PYR$_{14}$TFSI–LiTFSI). They prepared the polymer cl-PEO–PYR$_{14}$TFSI–LiTFSI with ILs and cross-linked poly(ethyleneoxide) and PIL–PYR$_{14}$TFSI–LiTFSI with pyrrolidinium-based polymeric IL. They used LTO and LFP electrodes using sodium carboxymethylcellulose (CMC), a water-soluble binder. Batteries with cl-PEO–PYR$_{14}$TFSI–LiTFSI as electrolyte and LFP as cathode and Li metal anode deliver nearly the nominal capacity at 0.05 and it is stable for more than 1,000 cycles carried out at 40° C. Batteries using (0.9PYR$_{14}$FSI–0.1LiTFSI) electrolyte, LTO-based anode, and LFP-based cathode deliver 85% of the nominal capacity at 0.05 and are stable for almost 1,000 cycles carried out at 20° C [28].

A double polymer network gel electrolyte (DPNGE) is prepared with poly (vinylidene fluoride-co-hexafluoropropylene) (PVDF-HFP) and branched acrylate as the matrix by UV curing. The IL N-methyl-N-propylpyrrolidine bis(trifluromethanesulfonyl) imide (PYR$_{13}$TFSI) acts as plasticizer to provide high I_c. LiNO$_3$ is used to generate a stable SEI layer. The Li/DPNGE/LiFePO$_4$ shows 153.7 mA h/g capacitance at 0.5° C with capacitance retention 92.7% over 500 cycles [29]. Double-network ionogel (DN ionogel) electrolyte can show better performance. DN ionogel can be prepared by trapping EMIMTFSI in double network of crosslinked poly IL and PVDFHFP. It shows high conductivity and wide ESW. After assembling into LiFePO$_4$/Li cell, the DN ionogel exhibits capacitance of 115.6 mA h/g at 4° C and 98% capacitance retention after 200 cycles [30]. ILs are not only used as electrolytes; they can potentially be used in electrodes also. Patil et al. reported to prepare a cathode consisting of multifunctional surface and redox active poly (ionic liquid) bearing catechol pendent and multiwalled carbon nanotubes [31]. Not only in electrolyte but in electrodes also, improvements are tried for their better performance and making them more environmental-friendly [32]. Instead of common inorganic electrodes, scientists are trying to make organic electrodes which can provide better performance like ellagic acid [33] and calix [6] quinone (C6Q) [34]. For a greener approach natural binder can also be used for cathode and anode preparation [35].

10.3 FUEL CELLS

Fuel cells are primarily energy convertors which transform chemical energy to electrical energy. As its name suggests the cell uses 'fuel' (often hydrogen) and an oxidizing agent (often oxygen) as chemical energy source and converts it into electrical energy with the help of some redox element and produces water and heat as by-product. Other than hydrogen, methanol, ethanol, ammonia, etc., are used as fuel. Unlike batteries it needs a continuous supply of fuel to work. The simplest fuel cell consists of two electrodes with an electrolyte in between. Hydrogen is fed to the anode which generates a proton and an electron. The proton travels through the electrolyte to the cathode where it meets with the oxygen provided and with the help of a redox element present in the cathode it converts into water. The electrolyte works as an insulator of electron, so the electrons travelled through an external electrical circuit recombine in the cathode. The reaction involved in the anode and cathode is given below.

Anode reaction: Cathode reaction:

Anode reaction: Cathode reaction :

$$H_2 \rightarrow 2H^+ + 2e^- \qquad \frac{1}{2}O_2 + 2H^+ + 2e^- \rightarrow H_2O$$

There are different types of fuel cells available. Fuel cells are classified mainly either according to the electrolyte used or the fuel fed. There are six types of fuel cells available: (i) proton exchange membrane fuel cell (PEMFC), (ii) alkaline fuel cell (AFC), (iii) phosphoric acid fuel cell (PAFC), (iv) molten carbonate fuel cell (MCFC), (v) solid oxide fuel cell (SOFC), and (vi) direct methanol fuel cell (DMFC). They can be further classified according to the working temperature [36]. Fuel cells have many advantages like in comparison to combustion engine can operate at high efficiency exceeding 60% or more, highly reliable, easy to install, and lower or zero emission.

PEMFCs: In PEM fuel cells a solid or semisolid polymer electrolyte membrane that exchange protons is used. PEMs are made of proton conducting acid polymer mostly perfluoro sulfonic acid polymer, and 'NAFION' or 'AQUNION' is the commercially used one. Proton exchange occurs through either vehicle mechanism (by forming complex H_3O^+, $H_5O_2^+$ or $H_9O_4^+$) or Grothuss mechanism (proton hopping from one to another carrier). At high-temperature vehicle method does not work and Grothuss method to work short distance (<0.275 nm) between carriers is needed [37]. Nafion-based PEMs form LT-PEMFCs. The degree of sulfonation increases the number of protons conducting ion [38,39]. But increase in the degree of sulfonation makes the polymer extremely hydrophilic which consecutively causes poor mechanical strength, big swelling, increased fuel permeability [38], and conductivity reduction above 100°C [40]. Also, nafion polymer is highly acidic due to the presence of sulfonic acid group that required such environment which could only be provided by noble metal catalysts, i.e., Pt. Use of noble metal catalyst increases the cost of PME fuel cells. Also, to operate properly the membrane needs to be hydrated. It also limits the operating temperature under 100°C. For HT-PEMFCs PBI is commercially used with PA as PBI offers superior thermal and mechanical stability [41]. But PBI/PA membrane has some drawbacks like partial loss of PA in lower temperature (100°C) and high price [42,43]. HT-PEMFCs furnish several advantages like simple water and heat management, amplify electrode kinetics, and high endurance of the CO in the fuel [44–46]. This demands further development in this sector and requires new polymer electrolyte membrane independent of hydration. In this case substitute polymers like poly (vinyl alcohol) [47], poly (ether-ether-ketone) [48], PI, poly(phenylene) [49], poly(sulfone), and poly (vinylidene fluoride) [50] are investigated. A membrane to be feasible for PMEFCs must obey certain requirements. Membrane should be chemical, electrochemical and thermally stable, must have mechanical stability, highly conductive of proton irrespective of humidity, low cost, and low oxygen and fuel crossover. ILs are great candidates in this case. PILs offer high

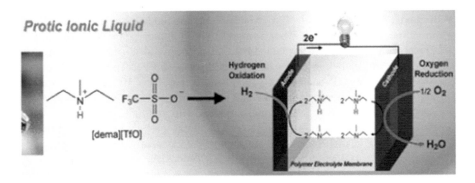

FIGURE 10.2 Proton exchange membrane fuel cell (PEMSC) with PIL [40]. (Reproduced with permission from Kanno et al. (Ref. 40). Copyright 2010 American Chemical Society.)

proton conduction, hydration independent, and high operative voltage. The IL incorporation in the polymer is a matter of approach and can be divided into further classes (Figure 10.2).

Supported IL membrane (SILM): SILMs consist of a liquid IL supported on a porous polymer membrane by capillary force. It provides the advantages like need for low solvent quantity, high surface area coverage, and high efficiency. A lot of researchers are trying to improve the nafion membrane. Nafion works as a connector between catalyst particle on electrode and polymer. But catalyst performance decreases due to the adsorption of sulfonate groups of nafion on the catalyst surface. To reduce this impact ILs can be incorporated. Addition of ILs on the nafion surface completely removes the negative impact of nafion on the ORR kinetics activity [51]. Main drawback of this type is the lack of durability as ILs tends to leak from the pores.

Polymer IL inclusion membrane (PILIM): PILIMs are formed by immobilizing the IL into the polymer matrix. It gives improved mechanical and chemical properties in comparison with SILMs. Malis et al. prepared membranes by casting two polymer Poly (vinylidene fluoride-co-hexafluoropropane) or fluoroeldstomer and nafion with ILs [BMIM][TFO] and [EIM][TFO]. The four combination shows significant I_c in both hydrated and anhydrous condition. The highest power contribution was 1.2 mW cm^{-2} for [BMIM][TFO]-nafion in dry condition [52]. A new type of membrane was prepared by crosslinked sulfonated-PEEK (SPEEK) and ethylene glycol (EG) in presence of varied amount of [BMIM][OTf] IL (30, 40, 50, 60, and 70 wt%). Proton conductivity of all the membrane with SPEEK (degree of sulfonation 70%–72%): EG 67:33 is in the span 10^{-3} S cm^{-1} in temperature range 30°C–140°C in dry condition [53]. A new type of polymer was prepared by polyimide matrimid (PIM) and 30–60 wt% of [BAIM][TFSI]. The membrane is thermally stable up to 377°C–397°C. To increase its performance PIM was crosslinked with Jaffamine. New membrane shows I_c 10^{-2} S cm^{-1} at 130°C [54].

Polymerized IL membrane (PyILM): PyILMs are made by polymerizing IL monomers. A membrane was arranged by copolymerizing [HSO_3-BVIm][TfO] with methyl methacrylate (MMA) and perfluoro-3,6-dioxa-4-methyl-7-octene sulfonyl fluoride in its hydrated form (HPFSVE). Poly (IL-co-MMA) and poly (IL-co-HPFSVE) shows I_c in span 10^{-2}–10^{-3} S cm^{-1} in both hydrated and anhydrous condition and utmost power output 45.76 and 28.12 mW cm^{-2}, respectively. Membranes are steady up to 200°C [55].

Polymer-IL composite membrane: Skorikova et al. prepared quasi-solidified membrane with PBI matrix and IL, then post-treatment with PA. Here they used two ILs [DEMA][NTf2] and [HHTMG][NTf2]. Two polymer membrane portrays conductivities in the range of 30–60 mS cm^{-1} at 180°C. [DEMA][NTf2]/PBI (1:1) provide highest power density of 0.32 W cm^{-2} at 200°C [56]. Another polymer membrane was made by incorporating [BMIM] IL with different anions (Cl$^-$, Br$^-$, I$^-$, NCS$^-$, NTf$_2^-$, PF$_6^-$, BF$_4^-$) in polymeric network of PBI. ILs upgrade the mechanical property of the membrane. I_c 94 mS cm^{-1} at 200°C in anhydrous condition was observed [57]. Lee et al. prepared a six-membered sulfonated polyimide/[DEMA][TfO] composite membrane. This membrane shows superior thermal stability, I_c, mechanical properties, and low gas permeability. This membrane works up to 140°C in anhydrous condition with current density 90 mA cm^{-2} and power density 19 mW cm^{-2} [58].

Inorganic–organic compound with PIL in membrane: Manufacture of inorganic–organic membrane by incorporation of inorganic nanoparticles such as SiO_2 and TiO_2 into nafion membrane is common approach to retain water molecule at high temperature [59,60]. Taking this approach in IL functionalized membrane, incorporation of inorganic-organic nanoparticles into PILs is a simple and effective way to immobilize ILs. But addition of right number of inorganic nanos are necessary as nano particles tends to aggregate in polymeric matrix. ILs can be mixed with various organic and inorganic materials. In this case Graphene Oxide (GO) is a good option. Lin et al. arranged poly(styrene-co-acrylonitrile)/PIL/functionalized GO membrane. GO was grifted with [APMIm][Br], and 1-methylimidazolium trifluoromethanesulfonate ([MIm][TfO] was used as PIL. A series of PIL-based PMEs were prepared with varied amount of [APMIm][Br]. All the PMEs show good thermal stability and high I_c. Membrane with 1.0 Wt% [APMIm][Br] gives maximum conductivity ~ 1.48×10^{-2} S cm^{-1} at 160°C. Even [APMIm][Br]-GO-based membrane shows better PIL detainment compared to membrane without [APMIm][Br] [60]. Zirconium hydrogen phosphate [$Zr(HPO_4)_2$] is also a good proton conductor. In previous studies Zr-IL material shows proton conduction 10^{-4} S cm^{-1} in anhydrous condition at 200°C. Then a new membrane was prepared using ZrP-IL, polytetrafluoroethylene (PTFE), and glycerol (GYL). They used 1-ethyl-3-methylimidazolium ethyl sulphate [EMIM][ESO4] as the IL. The highest proton conductivity with 0.36 mass fraction IL, 0.23 GLY and 0.41 ZrP was 0.0714 S cm^{-1}. At 200°C it shows anhydrous proton conductivity of 0.06 S cm^{-1} which is quite promising and equal to 60% of nafion at 80°C in fully hydrated condition [61]. But Zr compounds are costly enough. As an alternative Calcium Phosphate (CP) can be used. It has been reported before to be used in the preparation of fast proton conductors for electrochemical applications. In a recent study a CP-IL, GLY, PTFE membrane was prepared using two electrolyte 1-ethyl-3-methylimidazolium methanesulfonate ([EMIM][CH_3O_3S]) and 1-hexyl-3-methylimidazolium tricyanomethanide ([HMIM][C_4N_3]). The highest conductivity for [EMIM][CH_3O_3S] was found to be 5.01×10^{-3} S cm^{-1} for 0.6 wt% of IL and 0.6% of GLY at 25°C. That for 2.0% [HMIM][C_4N_3]/ 0.2% GLY/ CP/ PTFE at 200°C in anhydrous condition is 3.14×10^{-3} S cm^{-1} [62].

Not only in preparation of PEM electrolyte but also in preparation of electrode, ILs are used. Like Li prepared polymer electrolyte membrane coated electrode using Pt/Vulcan electrocatalyst loaded sulfonated poly (IL) block copolymer [63]. Biopolymers have also been recently studied to replace plastic polymers. Among them chitosan, cellulose, gellan gum, etc. [64] show good results. Biopolymers owns intrinsic thermochemical stability but has insufficient proton conductivity in the anhydrous state. Cellulose is a good choice of biopolymer as it's cheap, easy to access, and remains in its glassy state up to its deterioration temperature (200°C–220°C). A new membrane was prepared with cellulose nanocrystals (CNCs) and PILs. The two membranes prepared are CNC/Im (imidazole)/ [HC$_6$Im] [TfO] (N-hexylimidazolium trifluoromethylsulfonate) and CNC/Im/[HC$_6$Im][TFSI] (N-hexylimidazolium bis(trifluoromethylsulfonyl)imide). CNC/Im/PIL is thermally and chemical stable up to 200°C. The I_c is in the span of $4 \times 10^{-6} - 2 \times 10^{-3}$ S cm^{-1} between 30°C and 150°C in dry condition [65]. Kalaiselvimary et al. prepared a low-cost composite membrane, chitosan/PEO/H+ MMT (modified montmorillonite) which shows high water uptake and low swelling ratio [66]. Not only in preparation of PEMs but also as a binder for the electrode preparation, biopolymers are investigated. Ghobadi et al. prepared a graphene/cellulose/carbon fibre paper electrode via paper modelling process with ILs. After that Carbon supported Pt was sprayed for catalytic action. The highest power density was at 50% RH as 911 mW cm^{-2} [67]. Another electrode was prepared from Pt coated n doped carbon derived from aloe vera [68]. Some work is also done to replace the costly Pt with other materials [69,70].

Alkaline fuel cell (AFCs): AFCs are the most efficient fuel cell as it can reach 70% potential. Differing from PEM fuel cell, it uses aqueous alkali solution such as potassium hydroxide (KOH). The cell produces power through redox reaction also. Here in the anode H_2 is oxidized, whereas in cathode O_2 is reduced according to the following reactions.

Anode reaction:

$H_2 + 2OH^- \rightarrow 2H_2O + 2e^-$

Cathode reaction:

$O_2 + 2H_2O + 4e^- \rightarrow 4OH^-$

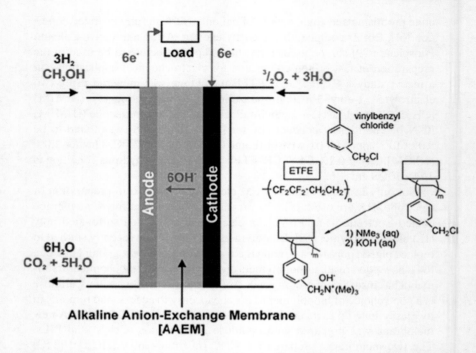

FIGURE 10.3 Working process of alkali anion exchange membrane fuel cell [72]. (Reproduced with permission from Yee et al. (Ref. 72). Copyright 2007 American Chemical Society.)

But AFC does not deny CO_2 so the fuel cell easily gets 'poisoned' by CO_2 which forms K_2CO_3 reacting with KOH. So, AFC needs to work in pure O_2 or at least filtered air. The main problem that poisoning creates is blockage of the electrode pores by deposition of carbonates which are less soluble at room temperature and depletion in the I_c of the electrolyte. AFCs works between room temperature to 90°C with better electrical efficiency than acidic fuel cell. As O_2 reduction on the cathode in alkali medium is more easy, non-noble metals can be used as catalysts. Iron, cobalt, or nickel can be used as anode catalyst, whereas silver or iron phalocyanines and spinels [71] can be used in cathodes. In HT no degradation in execution of the reaction is observed for normal AFCs because HT facilitates the solubility of K_2CO_3. But AFC suffers from high system level complexity due to the need for CO_2 purification and low CO_2 endurance. Alkali anion exchange membrane fuel cell (AAEMFCs) overcomes all the shortcomings of AFCs. Like PEM, AAEM is the membrane that can transport OH−. AAEMs offer benefits over PEMs. AAEMs generally contain positive ionic groups (typically quaternary ammonium) functional groups such as poly-N+Me$_3$ and immobile negative charged anions (Figure 10.3).

But alkyl quaternary ammonium group can be easily degraded at high temperature and PH by Hoffman elimination or nucleophilic substitution [73]. Imidazolium cations because of the large π-bonds are found to be more stable at elevated pH and temperature [74,75] and does not even form insoluble carbonates [76]. So new AEMs are being prepared by modifying commercially available polymers like polyether sulfone cardo, poly (phthalazinon ether sulfone ketone), polysulfone [77], poly styrene, poly phenylene [78], poly (acrylene ether sulfone). Fang et al. prepared copolymers using [VMIM][I] and [VBI][Br] with styrene. All membranes were found to be thermally stable. At 30°C the water uptake, IEC, and I_c of [VBI][Br]-styrene membrane was high as 56.8%, 1.26 m mol/g, and 2.26×10^{-2} S cm^{-1}, respectively. I_c increases with increasing temperature and the cell also shows good performance [79]. Other than ammonium or imidazolium, methylpyrrolidinium [80], cycloaliphatic quaternary ammonium [81], quaternary phosphonium [82], and guanidinium cations even organometallic cations such as cobaltocenium can also be used [83]. High IEC leads to excess swelling and thus OH− starts to attack polymer backbone. A new composite membrane was prepared with IL N, N, N′, N′-tetramethyl-1,6-hexanediamine and allyl bromide loading in hyper crosslinked polymer and then incorporation of it in quanternized poly(2,6-dimethyl-1,4-phenyleneoxide). It shows 92% enhancement in OH− conduction at IEC of 2.33 m mol g^{-1}. Even 88% of conductivity retention is observed under 2M NaOH solution [84]. An anion exchange membrane was prepared by linking pyridine and bipyridine in PEEK polymer. Addition of pyridine or bipyridine a strong passage of OH ion in the hydrophilic domain is formed. PYPEEK (36.99 mS cm^{-1}) generally shows higher I_c than BiPYPEEK (32.05 mS cm^{-1}) at 80°C. But BiPYPEEK shows better mechanical stability [85].

Like PEM cell inorganic nanoparticle layer addition like SiO_2, Al_2O_3 [86], ZrO_2 [87] also increases hydroxide ion conduction. Researchers prepared IL-coated silica in AEMs made of trimethylamine functionalized QAPPO. Two different ILs MIM and BPDMI were used. It shows good mechanical and thermal stability with better OH− conduction. QAPPO/8% MIM-SiO_2 shows low swelling ratio, OH− conduction of 32.3 and 70.2 mS cm^{-1} at 30°C and 80°C [88]. Not only low temperature but also high-temperature AEM can be produced. AEM made of low-density polyethylene (LDPE), where benzyltrimethylammonium (BTMA) head groups are covalently bonded, works in high temperature at 100°C. It offers power density of 2.1 W cm^{-2} and current density 574 mA cm^{-2} at 0.8V [89]. Another AEM was prepared which shows thermal decomposition at 300°C by copolymerizing bis-imidazolium functionalized norbornene IL with 5-norbornene-2-methylene glycidyl ether. It displays swelling only 20.86% and OH− conductivity 95×10^{-3} S cm^{-1} at 80°C [90].

Microbial fuel cells (MFCs): MFCs are a type of fuel cell that uses microorganisms to generate energy from chemicals. It mainly uses electron produced during microbial oxidation of fuel provided in anode and transfer it through external circuit to cathode where it reduces compounds. MFCs also have three components: anode, cathode and a membrane. But there is still need for a mediator that transfers electron from microbial cell to that electrode. The organisms consume substrates like sugar, acetate, and methane to produce CO_2, H^+, and electron according to the reaction $C_{12}H_{22}O_{11} + 13H_2O \rightarrow 12CO_2 + 48 H^+ + 48e^-$ in the absence of oxygen.

The mediator crosses the bacterial cell wall and taped in the electron transport chain. Then it takes up the electron that should have been taken by O_2 and exits the cell to transfer it to the anode. Releasing the electrons again transforms the mediator in its oxidizing state and thus a redox cycle goes on. As mediator methyl viologen, methylene blue, theonine, neutral red, humic acid, etc., are used. But most available mediators are expensive and toxic. Another option is to use those bacteria's that can transfer electron directly to the electrode through their electrochemically active proteins. The H^+ ion transfers from anode to cathode through a PME membrane. Generally, nafion is used. Koók et al. prepared a double-chambered MFC with a supported IL membrane using [BMIM][NTf$_2$] and [HMIM][PF$_6$] on PVDF layer. They used waste water of PH 7 and sodium acetate as fuel. Best energy yield was obtained for [BMIM][NTf$_2$] which is 18% higher than nafion 115 [91]. Where a polymer inclusion membrane was prepared based on [OMIM][NTf$_2$] and [MTOA]Cl ILs and PVC polymer. [MTOA]Cl with 70% W/W provides maximum power of 450 mW/m^3 and COD removal value 80%. Here increase in the amount of IL increases the power [92]. ILs can also be used in single-chambered MFC. When supported IL membrane using [C$_6$MIM][PF$_6$] and [BMIM][NTf$_2$] on PVDF was compared with nafion membrane with acetate, [BMIM][NTf$_2$] shows maximum energy output where [C$_6$MIM][PF$_6$] shows the least. But same with glucose, nafion shows the best performance [93]. Salar-García et al. prepared a polymer inclusion membrane using 70% W/W [P$_{14,14,14,1}$][TOS] and [MTOA]Cl ILS on PVC and THF as polymers and embedded them on carbon cloth. [P$_{14,14,14,1}$][TOS] offers higher maximum power output value of 12.3 W m^{-3} and COD removal 60% [94].

10.4 SOLAR CELLS

Solar cells have the power to change the future of energy generation. But there is still some barrier in commercializing them due to their unstable nature and cost. Solar cell or a photovoltaic cell is an energy convertor that converts photo energy to electrical energy. It works in several stages. First photon in the sunlight is absorbed by a semiconductor which excites their valence electron. This excited electron transports through an external circuit. Thus, direct current is produced. Commonly dopped silicon is used as a semiconductor. However recently an alternative source of energy-generating compounds is used. Now third-generation solar cells (like dye-sensitized solar cells, perovskite solar cell, and organic solar cell) are more popular.

Dye-sensitized solar cell (DSSCs): It is a photoelectrochemical system which is based on a semiconductor formed between two electrode layers and an electrolyte in between. It has attractive features like they are transparent/semi-transparent, can absorb defused light, and flexible. It is made up of two TCO layers among which one is attached to a layer of dye-sensitized particle absorbed titanium oxide nanoparticles and electrolyte filled in between.

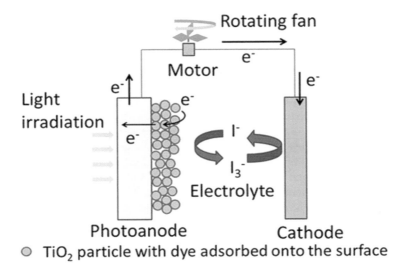

FIGURE 10.4 Schematic diagram of working procedure of DSSCs [96]. (Reproduced with permission from Chou et al. (Ref. 96). Copyright 2018 American Chemical Society.)

Here the dye-sensitized particle absorbs the photon energy which leads to photoexcitation of electron. This electron is ejected in the conduction band of TiO_2 through which it flows through external circuit. Then again through a redox reaction (mostly I^-/I_3^-) in electrolyte dye-sensitized molecule gets back its electron. Generally, an organic solvent, ACN is used. I^-/I_3^- with organic electrolyte provides the best efficiency of 11.9% [95] (Figure 10.4).

Except I^-/I_3^-, Co^{II}/Co^{III} complex and Cu^I/Cu^{II} [97,98] are also used. However, it is not ideal due to its low boiling point and high vapor pressure; other nitriles (bp > 100°C) are also used like valeronitrile, 3-methoxypropionitrile (MPN), N-methyl-pyrrolidone, ethylene carbonate, and methylene carbonate. The shortage of stability due to corrosion of electrode and ion conductivity with leakage of electrolyte is the major drawback of commercial DSSCs [99]. To become an electrolyte for dye-sensitized solar cell it must be highly conductive, should have insignificant visible light absorption, must have high redox potential and thermally as well as chemically stable. IL are a great option for DSSCs as they improve charge transport and thus cell efficiency. But among various ILs, imidazolium ILs portrays excellent performance. Different imidazolium ILs with variable alkyl chain length was employed as redox mediator along with LiI and I_2 in MPN. Efficiency of the cell was found to decrease with increasing chain length. Butyl imidazolium iodide ILs shows good efficiency (4.96% and 5.17%) with improved I_c [100]. High viscosity of the ILs is the barrier in dye regeneration process which reduces the cell efficiency and V_{OC} [101]. Researchers prepared an electrolyte using eutectic mixture of bicyclic 1,2,3-triazolium iodide and bicyclic 1,2,3-triazolium tricyanomethanides. bicyclic 1,2,3-triazolium cation forms novel low viscosity ILs which were tested as electrolyte. Four electrolytes were prepared where alkyl chain length on the cation vary from C_1 to C_4. Best performance was observed for C_2 which attribute to J_{sc} 12.68 mA/cm^{-2}, V_{OC} 657 mV, and PEC 6.00%. The cell retains its performance up to 90% after 1,000 hours [102]. Solid electrolytes also can be made for the use of DSSCs. Polyoxometalates (POMs) can be added as the anion in IL. A series of IL cations were tested with $Mo_{12}O_{40}^{3-}$. Among them 1-butylpyridinium (BPy) and trihexyltetradecylphosphonium ($P_{6,6,6,14}$) gives the most promising results [103]. Bhagavathiachari et al. prepared polymer electrolyte with PVDF–MPN and compared with neat IL and IL solution in MPN. IL in MPN (IM) and IL in PVDF–MPN (IP) shows good performance, V_{OC} 830 and 810 mV, J_{sc} 11.1 and 11.9 mA cm^{-2}, efficiency 6.6% and 6.9%, respectively [104]. Polymer membranes can be prepared by IL grifting in the polymer also. Like different mole ratio pf PolyIL, poly[vinylidene fluoride-co-hexafluoro propylene-co-vinylideneaminooxomethyl-1-butylimidazolium iodide] (PFII), was grifted in PVDF-HFP. Three polymer PFII-F (mole ratio of IMIL ¼), PFII-E (IMIL 1/8), PFII-S (IMIL 1/16) was soaked in liquid redox solution. Increase in molar ratio of IMIL increases the redox absorption, diffusion coefficient and I_c. PFII even shows good charge transfer property. Maximum cell efficiency was 9.26% [105]. Diffusion coefficient of solid or gel electrolytes are low which in turn reduces the overall cell PCE. Quasi electrolytes are great solution for all these problems as it provides long-term stability like solid electrolytes and good interfacial contact and high I_c as liquids. A. Lennert et al. prepared quasi electrolytes using ILs with iodide and SeCN anions and iodide-pseudo halogen redox couple. ILs with iodide anion shows better J_{sc} and V_{OC}. 1,3-dialkylimidazolium iodide $(n=6)/(I_2)_{0.75}/((SeCN)_2)_{0.25}$ gives efficiency of 7%–8% and efficiency retention of 80% after 1,000 hours [106].

Perovskite solar cells (PSCs): PSCs are more efficient than DSSCs. It uses perovskite film as the absorber, an organic–inorganic hybrid ABX_3. 'A' is an organic cation such as methylammonium ($CH_3NH_3^+$) and formamide (FA). 'B' is an inorganic cation such as Sn^{2+} and Pb^{2+}, and 'X' is a halide. To improve the electron conduction and collection, numerous compounds have been subsuming in the interlayers of active material and electrodes, called ETL [107]. It offers efficiency greater than 25%. A PSCs consists of electrode/HTO/ perovskite/ETL/TCL. The grain size of a perovskite film is very important as the crystal defect plays a crucial role in determining the overall efficiency of the cell as they result in unwanted charge recombination [108–110]. Defects can be decreased if the grain size could be increased [111]. Addition of poly (methyl methacrylate) [112], poly(styrene) [113], polymer [114], liquid crystal, ILs [115] can increase the grain size and hydrophobicity of the perovskite layer.

When ILs are added in perovskite layer it forms a layer on top of them by self-assembled method due to their hydrophobicity [119]. Pb^{2+} in perovskite interacts with the lone pair of the IL molecule which leads to formation of layer by keeping the alkyl chain length vertical. The cation of the IL binds with the surface site of the perovskite resulting decline in the degradation

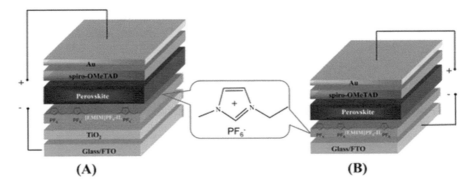

FIGURE 10.5 Schematic structure of PSC with (a) TiO_2 and IL, (b) IL without TiO_2 [122]. (Reproduced with permission from Liu et al (Ref. 122). Copyright 2016 American Chemical Society.)

and cation migration process. One of the problems in commercializing PSCs is its low stability in hydrated medium. Addition of different materials like metal halide, organic halide salts, inorganic salts, nanoparticles, fullerene, polymers, and ILs. can improve the stability of PSCs. Addition of $BMIMBF_4$ in triple cation perovskite $(FA_{0.83}MA_{0.17})_{0.95}Cs_{0.05}$ Pb $(I_{0.9}Br_{0.1})_3$ gives a spin coat on perovskite layer and significantly increase the grain size. The overall cell PCE is increased to 20.4% and gives 84% efficiency after 500 hours [120]. Another IL named 1-alkyl-4-amino-1,2,4-triazolium (RATZ), was added in $MAPbI_3$ which corresponds to PCE 20.03% and 80% efficiency retention in room temperature and dark condition in 40%±5% RH for 3,500 hours [121]. Xia et al. fabricated a $MAPbI_3$ perovskite layer with a novel π conjugated IL crystal (ILC) 4'-(N, N, N-trimethyl ammonium bromide hexyloxy)-4-cyanobiphenyl (6CNBP-N). The CN present in the ILC controls the crystal growth. The PCE upgrades to 20.45% wit ~80% retention of PCE after 20 days in RH 50%–60% [117]. Mainly TiO_2 is used as the ETL layer. The performance of TiO_2 can be improved by addition of various compounds. ILs are one of them. $EMIMPF_6$ was tested as an additive in TiO_2 layer and compared with only IL as independent ETL layer. PCE of PSC based on $MAPbI_3$ with TiO_2, IL with TiO_2 and only IL was 14.02% ± 0.43%, 18.42% ± 0.65% and 13.25% ± 0.55%, respectively [122] (Figure 10.5).

Organic solar cells (OSCs): PSCs use active layer made of organic polymers. The organic compounds used here has large π conjugated system. The cell mainly consists of an electron blocking layer of anode, indium tin oxide (ITO) glass substrate, active layer, that absorbs light and a metal electrode of Ca or Al as cathode. OSCs use PEDOT:PSS as HTL which gives work function (WF) of 5.1 eV. Other popular HTLs are derivatives of graphene, self-assembled small molecules, metal oxides such as ZnO, TiO_2 or water-soluble polymers like poly[(9,9-dioctyl-2,7-fluorene)-alt-(9,9-bis(3'-(N, N-dimethylamino)propyl)-2,7-fluorene)] (PFN) and polyethylenimine ethoxylated (PEIE). Ideal HTL must be highly conductive, easily processed and should be flexible. ILs can be a substantial additive to improve PEDOT:PSS performance. L. Huang et al. prepared IL functionalized PEDOT:PSS with PTB7:PC$_{71}$BM active compound, which displays J_{SC} 17.5 mA cm^{-2} and PCE 8.75%. Same with another active compound PTB7-Th:PC$_{71}$BM shows PCE 9.21%. Stability of the overall cell also increases [123].

Nonconjugated block polymer was prepared for use as ETL with hydrophobic polystyrene (PS) and hydrophilic polyIL 1-(4-vinylbenzyl)-3-methyl-imidazolium chloride (PSImCl), PS29-b-PSImCl$_x$ (where $x = 15$, 29, 45, and 60). As a result, WF of the ITO electrode decreases with increasing amount of x. hydrophobicity's of the ETL is more than PEIE. PS29-b-PSImCl$_{60}$ shows best performance with 7.55% PCE [124]. ILs can also be added as an extra layer ETL and cathode. Three ILs were investigated with different cation and same anion. The cell was prepared using PBDTTT-C:PC$_{71}$BM as active material. 1-butyl-3-methylimidazolium tetrafluoroborate (BMIM-BF$_4$) gives the poorest results, where tetrabutylammonium tetrafluoroborate (TBABF$_4$) and tetrabutylphosphonium tetrafluoroborate (TBPBF$_4$) gives good results. PEC was 7.29% and 7.16% for TEA and TBP, respectively [125]. In elevated temperature and humidity PEDOT:PSS manifested acidic property and hydrophilicity which results in the corrosion of the ITO cathode [126]. Cathode metal Al or Ca are also delicate in oxygen and humid environment [127]. So, inverted organic solar cells are formed. Here noble metal Cu, Au or Ag as cathode and ITO as anode is used, in which way PEDOT:PSS and non-noble metal can be avoided [128,129]. ETL is also important here because of its glassiness which effects the performance of the device. But to work ZnO or TiO$_2$ properly UV exposure is necessary [130]. However, SnO$_2$ can replace them. Tran et al. prepared a combined ETL with 1-benzyl-3-methylimidazolium chloride ([BzMIM]Cl) IL with SnO$_2$. This ETL in P3HT:PC60BM-based cell gives WF decrease from -4.38 eV to -3.82 eV and 81% PCE [131].

10.5 CONCLUSIONS

This chapter highlights the energy applications of ILs as energy storage and conversion materials and devices, explicitly focusing on the ILs applications as novel materials for fuel cells, LIBs, and solar cells. Looking at the current situation of continuously increasing demand for a sustainable society, scientists across the globe are putting their best efforts towards clean and sustainable energy to tackle the environmental problems and to make the transition to a fossil-fuel-free society. Owing to their unique physicochemical properties, such as low melting point and high boiling point, high thermal stability, nonvolatility, low flammability, high I_C, and high electrochemical window, ILs seem to meet the strong demand as robust materials for energy conversion and storage devices. We have highlighted the use of a variety of ILs with different structures showing enormous potential as unique materials for energy storage devices like solar cells, fuel cells, and batteries. Due to the limitless tunability of ILs (tuning of the cation and/or anion structures), there is huge scope to explore the structure–property relationships that will definitely support the development of a variety of novel ILs with enhanced thermal and electrochemical performance for batteries and solar cells. This chapter provides an organized overview of the developing energy applications of ILs. This will open up novel avenues for the potential applications of these unique materials in various energy sectors and will offer some thoughts on the emerging opportunities and challenges.

Name	Abbreviations	Name	Abbreviations
Ionic liquid	IL	Lithium-ion batteries	LIBs
Solid electrolyte interface	SEI	Aprotic ionic liquids	APILs
Protic ionic liquids	PIL	Electrochemical stability window	ESW
Lithium titanate	LTO	Lithium-iron phosphate	LFP
Polybenzimidazole	PBI	Phosphoric acid	PA
Polymer exchange membrane	PEM	Low temperature	LT
Supported IL membrane	SILM	Polymerized IL membrane	PyILM
Microbial fuel cells	MFCs	Proton exchange membrane fuel cell	PEMFC
Alkaline fuel cell	AFC	Phosphoric acid fuel cell	PAFC
Molten carbonate fuel cell	MCFC	Solid oxide fuel cell	SOFC
Direct methanol fuel cell	DMFC	Microbial fuel cells	MFCs
Dye-sensitized solar sell	DSSCs	Perovskite solar cells	PSCs
Organic solar cells	OSCs		
High temperature	HT	Poly(imide)	PI
Indium tin oxide	ITO	Hole transfer layer	HTL
Power conversion energy	PCE	Electron transport layer	ETL
Transparent conducting layer	TCL	Transparent conducting oxide	TCO
Acetonitrile	ACN	3-methoxypropionitrile	MPN
Ionic conductivity	IC	Open-circuit voltage	VOP
Short-circuit current density	JSC	Fill factor	FF

REFERENCES

1. Fergus. J.W. Recent developments in cathode materials for lithium ion batteries. *J. Power Sourc.* **2010**, 195, 939–954.
2. Zhang, J.; Yu, A. Nanostructured transition metal oxides as advanced anodes for lithium-ion batteries. *Sci. Bull.* **2015**, 60(9), 823–838.
3. Zhanga, X.; Kühnelb, R.S.; Hua, H.; Edera, D.; Balducci, A. Going nano with protic ionic liquids – The synthesis of carbon coated $Li_3V_2(PO_4)_3$ nanoparticles encapsulated in a carbon matrix for high power lithium-ion batteries. *Nano Energy*, **2015**, 12, 207–214.

4. Jin, Y.; Kneusels, N.J.H.; Marbella, L.E.; Martinez, E.C.; Magusin, P.C.M.M.; Wheatherup, R.S.; Jonsson, E.; Liu, T.; Paul, S.; Grey, C.P. Under fluoroethylene carbonate and vinylene carbonate based electrolytes for Si anodes in lithium ion batteries with NMR spectroscopy. *J. Am. Chem. Soc.* **2018**, 140(31), 9854–9867.
5. Bai, L.; Ghiassinejad, S.; Brassinne, J.; Fu, Y.; Wang, J.; Yang, H.; Vlad, A.; Minoia, A.; Lazzaroni, R.; Gohy, J.F. High salt-content plasticized flame-retardant polymer electrolytes. *ACS Appl. Mater. Interf.* **2021**, 13, 44844–44859.
6. Kim, J.H.; Pieczonka, N.P.; Yang, L. Challenges and approaches for high-voltage spinel lithium-ion batteries. *ChemPhysChem* **2014**, 15, 1940–1954.
7. Sun, X.G.; Dai, S. Electrochemical investigations of ionic liquids with vinylene carbonate for applications in rechargeable lithium ion batteries. *Electrochimica Acta* **2010**, 55, 4618–4626.
8. Lingua, G.; Falco, M.; Stettner, T.; Gerbaldi, C.; Balducci, A. Enabling safe and stable Li metal batteries with protic ionic liquid electrolytes and high voltage cathodes. *J. Power Sourc.* **2021**, 481, 228979.
9. Buqa, H.; Wursig, A.; Vetter, J.; Spahr, M.E.; Krumeich, F.; Novak, P. SEI film formation on highly crystalline graphitic materials in lithium-ion batteries. *J. Power Sourc.* **2006**, 153, 385–390.
10. Qi, H.; Ren, Y.; Guo, S.; Wang, Y.; Li, S.; Hu, Y.; Yan. F. High-voltage resistant ionic liquids for lithium-ion batteries. *ACS Appl. Mater. Interf.* **2020**, 12, 591–600.
11. Appetecchi, B.; Montaninoa, M.; Carewskaa, M.; Morenoa, M.; Alessandrini, F.; Passerini, S. Chemical–physical properties of bis(perfluoroalkylsulfonyl)imide-based ionic liquids Giovanni. *Electrochimica Acta* **2011**, 56, 1300–1307.
12. Appetecchi, G.B.; D'Annibale, A.; Santilli, C.; Genova, E.; Lombardo, L.; Navarra, M.A.; Panero, S. Novel functionalized ionic liquid with a sulfur atom in the aliphatic side chain of the pyrrolidinium cation. *Electrochem. Commun.* **2016**, 63, 26–29.
13. Matsumoto, H.; Kageyama, H.; and Miyazaki, Y. Room temperature ionic liquids based on small aliphatic ammonium cations and asymmetric amide anions. *Chem. Commun.* **2002**, 8, 1726.
14. Abraham, D.P.; Roth, E.P.; Kostecki, R.; McCarthy, K.; MacLaren, S.; Doughty, D.H. Diagnostic examination of thermally abused high-power lithium-ion cells Abrahama. *J. Power Sourc.* **2006**, 161, 648–657.
15. Simonetti, E.; Maresca, G.; Appetecchi, G.B.; Kim, G.-T.; Loeffler, N.; Passerini, S.; Towards Li(Ni$_{0.33}$Mn$_{0.33}$Co$_{0.33}$)O$_2$/graphite batteries with ionic liquidbased electrolytes. I. Electrodes' behavior in lithium half-cells. *J. Power Sourc.* **2016**, 331, 426–434.
16. Kim, G.T.; Kennedy, T.; Brandon, T.; Geaney, H.; Ryan, K.M.; Passerini, S.; Appetecchi, G.B. Behavior of germanium and silicon nanowire anodes with ionic liquid electrolytes. *ACS Nano* **2017**, 11(6), 5933–5943.
17. Ishikawa, M.; Sugimoto, T.; Kikuta, M.; Ishiko, E.; Kono, M. Pure ionic liquid electrolytes compatible with a graphitized carbon negative electrode in rechargeable lithium-ion batteries. *J. Power Sourc.* **2006**, 162, 658–662.
18. Gehrke, S.; Ray, P.; Stettner, T.; Balducci, A.; Kirchner, B. Water in protic ionic liquid electrolytes: From solvent separated ion pairs to water clusters. *ChemSusChem* **2021**, 14, 1–11.
19. Brutti, S.; Simonetti. E.; Francesco, M.D.; Sarra, A.; Paolone, A.; Palumbo, O.; Fantini, S.; Lin, R.; Falgayrat, A.; Choi, H.; Kuenzel, M.; Passerini, S.; Appetecchi, G.B. Ionic liquid electrolytes for high-voltage, lithium-ion batteries. *J. Power Sourc.* **2020**, 479, 228791.
20. Hayashi, K.; Nemoto, Y.; Akuto, K.; Sakurai, Y. Alkylated imidazolium salt electrolyte for lithium cells. *J. Power Sourc.* **2005**, 146, 689–692.
21. Yang, B.; Li, C.; Zhou, J.; Liu, J.; Zhang, Q. Pyrrolidinium-based ionic liquid electrolyte with organic additive and LiTFSI for high-safety lithium-ion batteries. *Electrochimica Acta* **2014**, 148, 39–45.

22. Menne, S.; Schroeder, M.; Vogl, T.; Balducci, A. Carbonaceous anodes for lithium-ion batteries in combination with protic ionic liquids-based electrolytes. *J. Power Sourc.* **2014**, 266, 208–212.
23. Menne, S.; Pires, J.; Anouti, M.; Balducci, A. Protic ionic liquids as electrolytes for lithium-ion batteries. *Electrochem. Commun.* **2013**, 31, 39–41.
24. Stettner, T.; Walter, F.C.; Balducci, A. Imidazolium-based protic ionic liquids as electrolytes for lithium-ion batteries. *Batteries & Supercaps* **2018**, 2(1), 55–59.
25. Vogla, T., Passerinia, S., Balducci, A. The impact of mixtures of protic ionic liquids on the operative temperature range of use of battery systems. *Electrochem. Commun.* **2017**, 78, 47–50.
26. Kima, H.; Ding, Y.; Kohl, P.A. LiSICON – Ionic liquid electrolyte for lithium ion battery. *J. Power Sourc.* **2012**, 198, 281–286.
27. Forgie, J.C.; Khakani, S.; MacNeila, D.D.; Rochefort, D. Electrochemical characterization of a lithium-ion battery electrolyte based on mixtures of carbonates with a ferrocene-functionalised imidazolium electroactive ionic liquid. *Phys. Chem. Chem. Phys.* **2013**, 15, 7713–7721.
28. Balducci, A.; Jeonga, S.S.; Kima, G.T.; Passerini, S.; Winter, M.; Schmuck, M.; Appetecchi, G.B.; Marcilla, R.; Mecerreyes, D.; Barsukove, V.; Khomenkoe, V.; Canterof, I.; De Meatzaf, I.; Holzapfel, M.; Tran, N. Development of safe, green and high performance ionic liquids-based batteries (ILLIBATT project). *J. Power Sourc.* **2011**, 196, 9719–9730.
29. Wu, Q.; Yang, Y.; Chen, Z.; Su, Q.T.; Huang, S.; Song, D.; Zhu, C.; Ma, R.; Li, C. Dendrite-free solid-state Li metal batteries enabled by bifunctional polymer gel electrolytes. *ACS Appl. Energy Mater* **2021**, 4, 9420–9430.
30. Liang, L.; Chen, X.; Yuan, W.; Chen, H.; Liao, H.; Zhang, Y. Highly conductive, flexible, and nonflammable double-network poly(ionic liquid)-based ionogel electrolyte for flexible lithium-ion batteries. *ACS Appl. Mater. Interf.* **2021**, 13, 25410–25420.
31. Patil, N.; Aqil, M.; Aqil, A.; Ouhib, F.; Marcilla, R.; Minoia, A.; Lazzaroni, R.; Jerôme, C.; Detrembleur, C. Integration of redox-active catechol pendants into poly(ionic liquid) for the design of high-performance lithium-ion battery cathodes. *Chem. Mater.* **2018**, 30, 5831–5835.
32. Lin, C.; Wang, G.; Lin, S.; Lia, J.; Lu, L. $TiNb_6O_{17}$: A new electrode material for lithium-ion batteries. *Chem. Commun* **2015**, 51(43), 8970–8973.
33. Goriparti, S.; Harish, M.N.K.; Sampath, S. Ellagic acid – A novel organic electrode material for high capacity lithium ion batteries. *Chem. Commun* **2013**, 49, 7234.
34. Zhang, X.; Zhou, W.; Zhang, M.; Yang, M.; Huang, W. Superior performance for lithium-ion battery with organic cathode and ionic liquid electrolyte. *J. Energy Chem.* **2021**, 52, 28–32.
35. Kim, G.T., Jeong, S.S., Joost, M., Rocca, E.; Winter, M.; Passerini, S.; Balducci, A. Use of natural binders and ionic liquid electrolytes for greener and safer lithium-ion batteries. *J. Power Sources* **2011**, 196, 2187–2194.
36. Kirubakaran, A.; Jain, S.; Nema, R.K. A review on fuel cell technologies and power electronic interface. *Rene. Sustain. Energy Rev.* **2009**, 13, 2430–2440.
37. Presiado, I.; Lal, J.; Mamontov, E.; Kolesnikov, A.I.; Huppert, D. Fast proton hopping detection in Ice Ih by quasi-elastic neutron scattering. *J. Phys. Chem. C* **2011**, 115, 10245–10251.
38. Roelofs, K.S.; Kampa, A.; Hirth, T.; Schiestel, T. Behavior of sulfonated poly(ether ether ketone) in ethanol–water systems. *J. Appl. Polym. Sci.* **2010**, 111(6), 2998–3009.
39. Wang, G.; Guiver, M.D. Proton exchange membranes derived from sulfonated polybenzothiazoles containing naphthalene units. *J. Membr. Sci.* **2017**, 542, 159–167.
40. Lee, S.-Y.; Ogawa, A.; Kanno, M.; Nakamoto, H.; Yasuda, T.; Watanabe, M. Nonhumidified intermediate temperature fuel cells using protic ionic liquids. *J. Am. Chem. Soc.* **2010**, 132, 9764–9773.

41. Araya, S.S.; Zhou, F.; Liso, V.; Sahlin, S.L.; Vang, J.R.; Thomas, S.; Gao, X.; Jeppesen, C.; Kaer, S.K. A comprehensive review of PBI-based high temperature PEM fuel cells. *Int. J. Hydrogen Energy* **2016**, 41, 21310–21344.
42. Li, Q.; Jensen, J.O.; Savinell, R.F.; Bjerrum, N.J.; High temperature proton exchange membranes based on polybenzimidazoles for fuel cells. *Prog. Polym. Sci.* **2009**, 34, 449–477.
43. Chandan, A.; Hattenberger, M.; El-kharouf, A.; Du, S.; Dhir, A.; Self, V.; Pollet, B.G.; Ingram, A.; Bujalski, W. High-temperature (HAT) polymer electrolyte membrane fuel cells (PEMFC) e a review. *J. Power Sourc.* **2013**, 231, 264–278.
44. Uregen, N.; Pehlivanoglu, K.; Ozdemir, Y.; Devrim, Y. Development of polybenzimidazole/graphene oxide composite membranes for high temperature PEM fuel cells. *Int. J. Hydrogen Energy* **2017**, 42, 2636–2647.
45. Quartarone, E.; Angioni, S.; Mustarelli, P. Polymer and composite membranes for proton conducting, high-temperature fuel cells: A critical review. *Materials* **2017**, 10, 687–703.
46. Dupuis, A.-C. Proton exchange membranes for fuel cells operated at medium temperatures: Materials and experimental techniques. *Prog. Mater. Sci.* **2011**, 56, 289–327.
47. Liew, C.-W.; Ramesh, S.; Arof, A.K. A novel approach on ionic liquid-based poly(vinyl alcohol) proton conductive polymer electrolytes for fuel cell applications. *Int. J. Hydrogen Energy* **2014**, 39(6), 2917–2928.
48. Zhang, H.; Wu, W.; Wang, J.; Zhang, T.; Shi, B.; Liu, J.; Cao, S Enhanced anhydrous proton conductivity of polymer electrolyte membrane enabled by facile Ionic liquid-based hoping pathways. *J. Mem. Sci.* **2015**, 476, 136–147.
49. Miyatake, K.; Iyotani, H.; Yamamoto, K.; Tsuchida, E. Synthesis of Poly(phenylene sulfide sulfonic acid) via Poly(sulfonium cation) as a thermostable proton conducting polymer. *Macromolecules* **1996**, 29, 6969–6971.
50. Fernández, I, V.; Raghibi, M.; Bouzina, A.; Timperman, L.; Bigarre, J.; Anouti, M. Protic Ionic liquids/poly(vinylidene fluoride) composite membranes for fuel cell application. *J. Energy Chem.* **2020**, 53, 194–207.
51. Li, Y.; Intikhab, S.; Malkani, A.; Xu, B.; Snyder, J.D. Ionic liquid additives for the mitigation of nafion specific adsorption on platinum. *ACS Catal.* **2020**, 10(14), 7691–7698.
52. Malis, J.; Mazúr, P.; Schauer, J.; Paidar, M.; Bouzek, K. Polymer-supported 1-butyl-3-methylimidazolium trifluoromethanesulfonate and 1-ethylimidazolium trifluoromethanesulfonate as electrolytes for the high temperature PEM-type fuel cell. *Int. J. Hydrogen Energy* **2013**, 38, 4697–4704.
53. Malik, R.S.; Verma, P.; Choudhary, V. A study of new anhydrous, conducting membranes based on composites of aprotic ionic liquid and cross-linked SPEEK for fuel cell application. *Electrochimica Acta* **2015**, 152, 352–359.
54. Rogalsky, S.; Bardeau, J.-F.; Makhno, S.; Tarasyuk, O.; Babkina, N.; Cherniavska, T.; Filonenko, M.; Fatyeyeva, K. New polymer electrolyte membrane for medium-temperature fuel cell applications based on cross-linked polyimide Matrimid and hydrophobic protic ionic liquid. *Mater. Today Chem.* **2021**, 20, 100453.
55. Ortiz-Martíneza, V.M.; Ortiza, A.; Fernández-Stefanutoc, V.; Tojoc, E.; Colpaertb, M.; Amédurib, B. Fuel cell electrolyte membranes based on copolymers of protic ionic liquid [HSO$_3$-BVIm][TfO] with MMA and hPFSVE, *Inmaculada Ortiz Poly.* **2019**, 179, 121583.
56. Skorikova, G.; Rauber, D.; Aili, D.; Martin, S.; Li, Q.; Henkensmeier, D.; Hempelmann, R. Protic ionic liquids immobilized in phosphoric acid-doped polybenzimidazole matrix enable polymer electrolyte fuel cell operation at 200°C. *J. Mem. Sci.* **2020**, 608, 118188.
57. Escorihuela, J.; García-Bernabé, A.; Montero, Á.; Sahuquillo, Ó.; Giménez, E.; Compañ, v. Ionic liquid composite polybenzimidazol membranes for high temperature PEMFC Applications, *Polymers* **2019**, 11, 732.

58. Aricò, A.S.; Baglio, V.; Blasi, A.D.; Creti, P.; Antonucci, P.L; Antonucci, V. Influence of the acid–base characteristics of inorganic fillers on the high temperature performance of composite membranes in direct methanol fuel cells. *Solid State Ionics* **2003**, 161(3–4), 251–265.
59. Zhou, Y.; Xiang, W.; Chen, S.; Fang, S.; Zhou, X.; Zhang, J.; Lin, Y. Improvements of photocurrent by using modified SiO_2 in the poly(ether urethane)/poly(ethylene oxide) polymer electrolyte for all solid-state dye-sensitized solar cells. *Chem. Commun.* **2009**, 26(26), 3895.
60. Lin, B.; Yuan, W.; Xu, F.; Chen, Q.; Zhu, H.; Li, X.; Yuan, N.; Chu, F.; Ding, J. Protic ionic liquid/functionalized graphene oxide hybrid membranes for high temperature proton exchange membrane fuel cell applications. *Appl. Surf. Sci.* **2018**, 455, 295–301.
61. Al-Othman, A.; Nancarrow, P.; Tawalbeh, M.; Ka'ki, A.; El-Ahwal, K.; El Taher, B.; Alkasrawi, M. Novel composite membrane based on zirconium phosphate-ionic liquids for high temperature PEM fuel cells. *Int. J. Hydrogen Energy* **2021**, 46(8), 6100–6109.
62. Ka'ki, A.; Alraeesi, A.; Al-Othman, A.; Tawalbeh, M. Proton conduction of novel calcium phosphate nanocomposite membranes for high temperature PEM fuel cells applications. *Int. J. Hydrogen Energy* **2021**, 46(59), 30641–30657.
63. Li, Y.; Cleve, T.V.; Sun, R.; Gawas, R.; Wang, G.; Tang, M.; Elabd, Y.A.; Snyder, J.; Neyerlin, K.C. Modifying the electrocatalyst–ionomer interface via sulfonated poly(ionic liquid) block copolymers to enable high performance polymer electrolyte fuel cells. *ACS Energy Lett.* **2020**, 5, 1726–1731.
64. Naachiyar, R.M.; Ragam, M.; Selvasekarapandian, S.; Krishna, M.V.; Buvaneshwari, P. Development of biopolymer electrolyte membrane using Gellan gum biopolymer incorporated with NH_4SCN for electro-chemical application. *Ionics* **2021**, 3415–3429.
65. Danyliv, O.; Strach, M.; Nechyporchuk, O.; Nypelö, T.; Martinelli, A. Self-standing, robust membranes made of cellulose nanocrystals (CNCs) and a protic ionic liquid: Toward sustainable electrolytes for fuel cells. *ACS Appl. Energy Mater.* **2021**, 4, 6474–6485.
66. Kalaiselvimary, J.; Selvakumar, K.; Rajendran, S.; Sowmya, G.; Ramesh Prabhu, M. Effect of surface-modified montmorillonite incorporated biopolymer membranes for PEM fuel cell applications. *Poly. Comp.* **2017**, 40(S1), E301–311.
67. Ghobadi, S.; Şanlı, L.I.; Bakhtiari, R.; Gürsel, S.A, Green composite papers via use of natural binders and graphene for PEM fuel cell electrodes. *ACS Sustain. Chem. Eng.* **2017**, 5, 8407–8415.
68. Varathan, P.; Akula, S.; Moni, P.; Sahu, A.k. Natural aloe vera derived Pt supported N-doped porous carbon: A highly durable cathode catalyst of PEM fuel cell. *Int. J. Hydrogen Energy* **2020**, 45(38), 19267–19279.
69. Wang, M.; Zhang, H.; Thirunavukkarasu, G.; Salam, I.; Varcoe, J.R.; Mardle, P.; Li, X.; Mu, S.; Du, S. Ionic liquid-modified microporous ZnCoNC based electrocatalysts for polymer electrolyte, fuel cells. *ACS Energy Lett.* **2019**, 4, 2104–2110.
70. Brkovic, A.M.; Kaninski, M.P.M.; Lausevic, P.Z.; Saponjic, A.B.; Radulovic, A.M.; Rakic, A.A.; Pasti, I.A.; Nikolic, V.M. Non-stoichiometric tungsten-carbide-oxidesupported PtRu anode catalysts for PEM fuel cells - From basic electrochemistry to fuel cell performance. *Int. J. Hydrogen Energy* **2020**, 45(27), 13929–13938.
71. Ho, J.; Li, Y.; Dai, Y.; Kim, T.; Wang, J.; Ren, J.; Yun, H.; Liu, X. Ionothermal synthesis of N-doped carbon supported $CoMn_2O_4$ nanoparticles as ORR catalyst in direct glucose alkaline fuel cell. *Int. J. Hydrogen Energy* **2021**, 46, 20503–20515.
72. Varcoe, J.R.; Slade, R.C.T.; Yee, E.L.H.; Poynton, S.D.; Driscoll, D.J.; Apperley, D.C. Poly(ethylene-co-tetrafluoroethylene)-derived radiation-grafted anion-exchange membrane with properties specifically tailored for application in metal-cation-free alkaline polymer electrolyte fuel cells. *Appl. Chem. Mater.* **2007**, 19, 2686–2693.

73. Varcoe, J.R.; Slade, R.C.T.; Prospects for alkaline anion-exchange membranes in low temperature fuel cells. *Fuel Cells* **2005**, 5, 187–200.
74. Ngo, H.L.; LeCompte, K.; Hargens, L.; McEwen, A.B. Thermal properties of imidazolium ionic liquids. *Thermochim. Acta* **2000**, 357, 97–102.
75. Ye, Y.; Elabd, Y.A. Relative chemical stability of imidazolium-based alkaline anion exchange polymerized ionic liquids. *Macromolecules* **2011**, 44, 8494–8503.
76. Cadena, C.; Anthony, J.L.; Shah, J.K.; Morrow, T.I.; Brennecke, J.F.; Maginn, E.J. Why is CO_2 so soluble in imidazolium-based ionic liquids? *J. Am. Chem. Soc.* **2004**, 126(16), 5300–5308.
77. Urenãa, N.; Perez-Prior, M.T.; Rio, C.d.; Vareź, Levenfeld, B. New amphiphilic semi-interpenetrating networks based on polysulfone for anion-exchange membrane fuel cells with improved alkaline and mechanical stabilities. *Polymer* **2021**, 226, 123824.
78. Mondal, A.N.; Hou, J.; He, Y.; Wu, L.; Ge, L.; Xu, T. Preparation of click-driven cross-linked anion exchange membranes with low water uptake. *Particuology* **2020**, 48, 65–73.
79. Fang, J.; Lyu, M.; Wang, X.; Wu, Y.; Zhao, J. Synthesis and performance of novel anion exchange membranes based on imidazolium ionic liquids for alkaline fuel cell applications. *J. Power Sourc.* **2015**, 284, 517–523.
80. Meek, K.M.; Sun, R.; Willis, C.; Elabd, Y.A. Hydroxide conducting polymerized ionic liquid pentablock terpolymer anion exchange membranes with methylpyrrolidinium cations. *Polymer* **2017**, 134, 221–226.
81. Zhang, Y.; Chen, W.; Li, T.; Yan, X.; Zhang, F.; Wang, X.; Wu, X.; Pang, B.; He, G. Tuning hydrogen bond and flexibility of Nspirocyclic cationic spacer for high performance anion exchange membranes. *J. Membr. Sci.* **2020**, 613, 118507.
82. Wang, C.; Tao, Z.; Zhao, X.; Li, J.; Ren, Q. Poly(aryl ether nitrile)s containing flexible side-chain-type quaternary phosphonium cations as anion exchange membranes. *Sci. China Mater.* **2020**, 63, 533–543.
83. Yuan, H.; Liu, Y.; Tsai, T.-H.; Liu, X.; Kim, S.B.; Gupta, R.; Zhang, W.; Ertem, S.P.; Seifert, S.; Herring, A.M.; Coughlin, E.B. Ring-opening metathesis polymerization of cobaltocenium derivative to prepare anion exchange membrane with high ionic conductivity. *Polyhedron* **2020**, 181, 114462.
84. Li, P.; Li, K.; Chen, J.; Zhang, N.; Tang, S. Novel anion exchange membrane with poly ionic liquid-confined hypercrosslinked polymer for enhanced anion conduction and stability. *Int. J. Hydrogen Energy*, **2021**, 46(41), 21590–21599.
85. Yuan, Y.; Zhang, T.; Wang, Z. Preparation of an anion exchange membrane by pyridine-functionalized polyether ether ketone to improve alkali resistance stability for an alkali fuel cell. *Energy Fuels* **2021**, 35, 3360–3367.
86. Yang, C.C.; Chui, S.J.; Chein, W.C.; Chui, S.S. Quaternized poly(vinyl alcohol)/alumina composite polymer membranes for alkaline direct methanol fuel cells. *J. Power Sourc.* **2010**, 195, 2212–2219.
87. Li, X.H.; Yu, Y.F.; Meng, Y.D. Novel quaternized poly(arylene ether sulfone)/nano-ZrO_2 composite anion exchange membranes for alkaline fuel cells. *ACS Appl. Mater. Interf.* **2013**, 5, 1414–142.
88. Chen, N.; Liu, Y.; Long, C.; Li, R.; Wang, F.; Zhu, H. Enhanced performance of ionic-liquid-coated silica/quaternized poly(2,6- dimethyl-1,4-phenylene oxide) composite membrane for anion exchange membrane fuel cells. *Electrochimica Acta* **2017**, 258(20), 124–133.
89. Douglin, J.C.; Varcoe, J.R.; Dekel, D.R. A high-temperature anion-exchange membrane fuel cell. *J. Power Sourc. Adv.* **2020**, 5, 100023.
90. Zhang, F.; He, X.; Cheng, C.; Huang, S.; Duan, Y.; Zhu, C.; Guo, Y.; Wang, K.; Chen, D. Bis-imidazolium functionalized self-crosslinking block polynorbornene anion exchange membrane. *Int. J. Hydrogen Energy* **2020**, 45, 13090–13100.

91. Koók, L.; Nemestóthy, N.; Bakonyi, P.; Göllei, A.; Rózsenberszki, T.; Takács, P.; Salekovics, A.; Kumar, G.; Bélafi-Bakó, K. On the efficiency of dual-chamber biocatalytic electrochemical cells applying membrane separators prepared with imidazolium-type ionic liquids containing [NTf$_2$] - and [PF$_6$] – anions. *Chem. Eng. J.* **2017**, 324, 296–302.
92. Hernández-Fernández, F.J.; de los Ríos, A.P.; Mateo-Ramirez, F.; Juarez, M.D.; Lozano-Blanco, L.J.; Godinez, C. New application of polymer inclusion membrane based on ionic liquids as proton exchange membrane in microbial fuel cell. *Separat. Purif. Technol.* **2016**, 160, 51–58.
93. Koók, L.; Nemestóthy, N.; Bakonyi, P.; Zhen, G.; Kumar, G.; Lu, X.; Su, Saratale, G.D.; Kim, S.-H.; Gubicza, L. Performance evaluation of microbial electrochemical systems operated with Nafion and supported ionic liquid membranes. *Chemophere* **2017**, 175, 350–355.
94. Salar-García, M.J.; Ortiz-Martínez, V.M.; Baicha, Z.; de los Ríosc, A.P.; Hernandez-Fernández, F.J. Scaled-up continuous up-flow microbial fuel cell based on novel embedded ionic liquid-type membrane-cathode assembly. *Energy* **2016**, 101, 113–120.
95. Green, M.A.; Emery, K.; Hishikawa, Y.; Warta, W.; Dunlop, E.D. Solar cell efficiency tables. (Version 47). *Prog. Photovolt. Res. Appl.* **2016**, 24, 3–11.
96. Chien, S.-I., Su. C.; Chou, C.C.; Li, W.R. Visual observation and practical application of dye sensitized solar cells in high school energy education. *J. Chem. Educ.* **2018**, 95, 1167–1172.
97. Higashino, T.; Iiyama, H.; Nimura, S.; Kurumisawa, Y.; Imahori, H. Effect of ligand structures of copper redox shuttles on photovoltaic performance of dye-sensitized solar cells. *Inorg. Chem.* **2020**, 59, 452−459.
98. Liu, Y.; Yiu, S.-C.; Ho, C.-L.; Wong, W.-Y. Recent advances in copper complexes for electrical/light energy conversion. *Coord. Chem. Rev.* **2018**, 375, 514−557.
99. Wolfbauer, G.; Bond, A.M.; Eklund, J.C.; MacFarlane, D.R. A channel flow cell system specifically designed to test the efficiency of redox shuttles in dye sensitized solar cells. *Sol. Energ. Mat. Sol. C* **2001**, 70, 85–101.
100. Denizalti, S.; Ali, A.K.; Ela, Ç.; Ekmekci, M.; Erten-Ela, S. Dye-sensitized solar cells using ionic liquids as redox mediator. *Chem. Phys. Lett.* **2018**, 691, 373–378.
101. Matsumoto, H.; Matsuda, T.; Tsuda, T.; Hagiwara, R.; Ito, Y.; Miyazaki, Y. The application of room temperature molten salt with low viscosity to the electrolyte for dyesensitized solar cell. *Chem. Lett.* **2001**, 30, 26–27.
102. Shan Lau, G.P.; Tsao, H.N.; Zakeeruddin, S.M.; Grätzel, M.; Dyson, P.J. Highly stable dye-sensitized solar cells based on novel 1,2,3-triazolium ionic liquids. *ACS Appl. Mater. Interf.* **2014**, 6 (16), 13571–13577.
103. Cruz, H.; Pinto, A.N.; Lima, J.C.; Branco, L.C. Application of polyoxometalate-ionic liquids (POM-ILs) in dye-sensitized solar cells (DSSCs). *Sandra Gago Mat. Lett: X* **2020**, 6, 100033.
104. Bhagavathiachari, M.; Elumalai, V.; Gao, J.; Kloo, L. Polymer-doped molten salt mixtures as a new concept for electrolyte systems in dye-sensitized solar cells. *ACS Omega* **2017**, 2, 6570−6575.
105. Pang, H.W.; Yu, H.F.; Huang, Y.J.; Lia, C.T.; Ho, K.C. Electrospun membranes of imidazole-grafted PVDF-HFP polymeric ionic liquids for highly efficient quasi-solid-state dye-sensitized solar cells. *J. Mater. Chem. A*, **2018**, 6, 14215.
106. Lennert, A.; Sternberg, M.; Meyer, Costa, R.D.; Guldi, D.M. Iodine-pseudohalogen ionic liquid-based electrolytes for quasisolid-state dye-sensitized solar cells. *ACS Appl. Mater. Interf.* **2017**, 9(39), 33437–33445.
107. Li, T.; Chen, Z.; Wang, Y.; Tu, J.; Deng, X.; Li, Q. Materials for interfaces in organic solar cells and photodetectors. *ACS Appl. Mater. Interf.* **2020**, 12(3), 3301–3326.
108. Ran, C.; Xu, J.; Gao, W.; Huang, C., Dou, S. Defects in metal triiodide perovskite materials towards high-performance solar cells: Origin, impact, characterization, and engineering. *Chem. Soc. Rev.* **2018**, 47, 4581–4610.

109. Jain, B.S.M.; Phuyal, D.; Davies, M.L.; Li, M.; Philippe, B.; Castro, C.D.; Qiu, Z.; Kim, J.; Watson, T.; Tsoi, W.C.; Karis, O.; Rensmo, H.; Boschloo, G.; Edvinsson, T.; Durrant, J.R. An effective approach of vapour assisted morphological tailoring for reducing metal defect sites in lead-free, (CH3NH3)3Bi2I9 bismuth-based perovskite solar cells for improved performance and long-term stability. *Nano Energy* **2018**, 49, 614–624.
110. Azpiroz, J.M.; Mosconi, E.; Bisquert, J.; Angelis, F. De. Defect migration in methylammonium lead iodide and its role in perovskite solar cell operation. *Energy Environ. Sci.* **2015**, 8, 2118.
111. Olyaeefar, B.; Kandjani, S.A.; Asgari, A. Classical modelling of grain size and boundary effects in polycrystalline perovskite solar cells. *Sol. Energy Mater. Sol. Cell.* **2018**, 180, 76–82.
112. Wang, F.; Shimazaki, A.; Yang, F.; Kanahashi, K.; Matsuki, K.; Miyauchi, Y.; Takenobu, T.; Wakamiya, A.; Murata, Y.; Matsuda, K. Highly efficient and stable perovskite solar cells by interfacial engineering using solution-processed polymer layer. *J. Phys. Chem. C* **2017**, 121(3), 1562–1568.
113. Qin, P.-L.; Yang, G.; Ren, Z.-W.; Cheung, S. H.; So, S.K.; Chen, L.; Hao, J.; Hou, J.; Li, G. Stable and efficient organo-metal halide hybrid perovskite solar cells via π-conjugated Lewis base polymer induced trap passivation and charge extraction. *Adv. Mater.* **2018**, 30, 1706126.
114. Arivunithi, V.M.; Reddy, S.S.; Sree, V.G.; Park, H.-Y.; Park, J.; Kang, Y.-C.; Shin, E.-S.; Noh, Y.-Y.; Song, M.; Jin, S.-H. Introducing an organic hole transporting material as a bilayer to improve the efficiency and stability of perovskite solar cells. *Macromol. Res.* **2021**, 29, 149–156.
115. Bai, S.; Da, P.; Li, C.; Wang, Z.; Yuan, Z.; Fu, F.; Kawecki, M.; Liu, X.; Sakai, N.; Wang, J.T.W.; Huettner, S.; Buecheler, S.; Fahlman, M.; Gao1, F.; Snaith, H.J. Planar perovskite solar cells with long-term stability using ionic liquid additives. *Nature Lett.* **2019**, 571, 245–250.
116. Yamada, Y.; Nakamura, T.; Endo, M.; Wakamiya, A.; Kanemitsu, Y. Photocarrier recombination dynamics in perovskite $CH_3NH_3PbI_3$ for solar cell applications. *J. Am. Chem. Soc.* **2014**, 136, 11610–11613.
117. Xia, X.; Peng, J.; Wan, Q.; Zhao, J.; Wang, X.; Fan, Z.; Li, F. Functionalized ionic liquid-crystal additive for perovskite solar cells with high efficiency and excellent moisture stability. *ACS Appl. Mater. Interf.* **2021**, 13, 17677–17689.
118. Zhang, W.; Liu, X.; He, B.; Gong, Z.; Zhu, J.; Ding, Y.; Chen, H.; Tang, Q. Interface engineering of imidazolium ionic liquids toward efficient and stable $CsPbBr_3$ perovskite solar cells. *ACS Appl. Mater. Interf.* **2020**, 12, 4540–4548.
119. Wang, S.; Li, Z.; Zhang, Y.; Liu, X.; Han, J.; Li, X.; Liu, Z.; Liu, S.; Choy, W.C.H. Water soluble triazolium ionic-liquid-induced surface self-assembly to enhance the stability and efficiency of perovskite solar cells. *Adv. Funct. Mater.* **2019**, 29, 1900417.
120. Ali, N.; Liang, C.; Ji, C.; Zhang, H.; Sun, M.; Li, D. Fangtian Y., Zhiqun H. Enlarging crystal grains with ionic liquid to enhance the performance of perovskite solar cells. *Organ. Electron.* **2020**, 84, 105805.
121. Wang, S.; Li, Z.; Zhang, Y.; Liu, X.; Han, J.; Li, X.; Liu, Z.; Liu, S.; Choy, W.C.H. Water-soluble triazolium ionic-liquid-induced surface self-assembly to enhance the stability and efficiency of perovskite solar cells. *Adv. Funct. Mater.* **2019**, 1900417.
122. Wu, Q.; Zhou, W.; Liu, Q.; Zhou, P.; Chen, T.; Lu, Y.; Qiao, Q.; Yang, S. Solution-processable ionic liquid as an independent or modifying electron transport layer for high-efficiency perovskite solar cells. *ACS Appl. Mater. Interf.* **2016**, 8(50), 34464–34473.
123. Huang, L.; Cheng, X.; Zhang, L.; Zhou, W.; Xiao, S.; Tan, L.; Chen, L.; Chen, Y. High performance polymer solar cells realized by regulating the surface properties of PEDOT:PSS interlayer from ionic liquids. *ACS Appl. Mater. Interf.* **2016**, 8(40), 27018–27025.

124. Hwang, D.; Mecerreyes, D.; Jung, I.; Park, H.; Isik, M.; Park, J. Polystyrene-block-poly(ionic liquid) copolymers as work function modifiers in inverted organic photovoltaic cells. *ACS Appl. Mater. Interf.* **2018**, 10(5), 4887–4894.
125. Yu, W.; Zhou, L.; Yu, S.; Fu, P.; Guo, X.; Li, C. Ionic liquids with variable cations as cathode interlayer for conventional polymer solar cells. *Organ. Electron.* **2017**, 42, 387–392.
126. Jong, M.P.; Ijzendoorn, L.J.; Voigt, M.J.A. Stability of the interface between indium-tin-oxide and poly(3,4-ethylenedioxythiophene)/poly (styrenesulfonate) in polymer light-emitting diodes. *Appl. Phys. Lett.* **2000**, 77, 2255–2257.
127. Jørgensen, M.; Norrman, K.; Krebs, F.C. Stability/degradation of polymer solar cells. *Sol. Energy Mater. Sol. Cells* **2008**, 92, 686–714.
128. Zhang, F.; Xu, X.; Tang, W.; Zhang, J.; Zhuo, Z.; Wang, J.; Wang, J.; Xu, Z.; Wang, Y. Recent development of the inverted configuration organic solar cells. *Sol. Energy Mater. Sol. Cells* **2011**, 95, 1785–1799.
129. Wang, K.; Liu, C.; Meng, T.; Yi, C.; Gong, X. Inverted organic photovoltaic cells. *Chem. Soc. Rev.* **2016**, 45, 2937–2975.
130. Kim, J.; Kim, G.; Choi, Y.; Lee, J.; Park, S. Heum.; Lee, K. Light-soaking issue in polymer solar cells: Photoinduced energy level alignment at the sol-gel processed metal oxide and indium tin oxide interface. *J. Appl. Phys.* **2012**, 111, 114511-1–114511-9.
131. Trana, V.; Khana, R.; Lee, I.; Lee, S. Low-temperature solution-processed ionic liquid modified SnO_2 as an excellent electron transport layer for inverted organic solar cells. *Sol. Energy Mater. Sol. Cells* **2018**, 179, 260–269.

Index

absorption 60–62, 68, 69–70, 71
adiabatic 83
amorphous 59–61, 73
anionic doping 5, 12
anode 173
ANSYS FLUENT 82, 83, 84, 85, 96, 139, 140, 141, 168, 169

band edges 7
bandgap 43, 57, 59, 60–73, 75
BGK approximation 166
binary 57, 65–67, 73–74
bounce back 166
boundary conditions 144

carbon capture 186
carbon dioxide reduction reaction 9
cathode 173
cationic doping 11, 14
chalcopyrite 63, 71–72, 75
charge transfer mechanism 18
chemical vapor deposition 174
circulation 162, 163, 164, 165
co-catalyst 15
computational fluid dynamics (CFD) 138, 139, 140
conductivity 187, 188, 189, 190
convergence study 147

D2Q9 lattice 167
density distribution 148
deep learning 125, 130
deposition 61–64
DFT 67–70, 74–75
diamond 95
direct absorber solar collectors (DASCs) 80
discrete lattice velocity 166
doping 11
   anion 5, 11
   cation 11, 13
dust deposition 29
   causes of dust deposition 30
   effect on the PV system 29–30
   mitigation techniques 30–32

effective mass 6–7
electrochemical window 186, 187, 188, 189
electronic properties 67–68
energy density 172
energy storage 117, 118, 119, 120, 121, 122, 123, 129, 130, 131, 172, 186

ETC 103–107, 111
evacuated tube solar water heater 101–103, 105, 107, 109, 111, 112
experimental setup 101, 106, 107, 112

fill factor 48
flow rate 101, 104–106, 108–112
fossil fuels 171, 185
fuel cell 134, 135, 137, 186, 189
   alkaline fuel cell 190, 193
   direct methanol fuel cell 190
   inorganic–organic compound with PIL in membrane 192
   microbial fuel cell 195
   molten carbonate fuel cell 190
   nafion 190, 191, 192
   phosphoric acid fuel cell 190
   polymer-IL composite membrane 192
   polymer IL inclusion membrane 191
   polymerized IL membrane 192
   proton exchange membrane 190, 192, 194, 196
   proton exchange membrane fuel cell 190, 193
   solid oxide fuel cell 190
   supported IL membrane 191
functional 57, 67–68, 70
functionalization 11

geometric model 144
Gibbs free energy 10
graphene 119, 120, 121, 122, 123, 129, 131, 132
graphite 186, 187
green chemistry 186
green technology 120, 121, 123, 125, 127

hailstorm 32
   effect on the PV system 32
heat pipe 102, 103
heterojunction 15
heterostructure 15
high-pressure 102, 103
hydrogen evolution reaction 7–8
hydrogen gas 150, 151, 152, 153, 154, 155, 156, 157

intermediate 57, 70–73, 75–76
ionic liquids 186, 187, 188, 189, 190, 193, 196, 197, 198, 199, 200
   alkylammonium 186, 187, 195
   alkyl phosphonium 186, 195
   alkyl pyridinium 186, 187, 189

211

ionic liquids (*cont.*)
   aprotic ionic liquid 188
   N-alkyl imidazolium 186, 187, 195
   protic ionic liquid 188
   pyrrolidinium 186, 187, 195
IoT based PV cleaning technique 33–40
isotherms 87, 88, 89, 90, 91

Lattice Boltzmann method (LBM) 165, 166
lithium-ion battery 186, 187, 188

machine learning 117, 124, 125, 128, 129, 130
mass conservation 141
material 103, 105–107
maximum efficiency 104, 105, 111, 112
Maxwell–Garnet relation 84
mechanical exfoliation 174
membrane wall properties 144
mesh 145, 148
mesh independence 85
modernization 101
momentum conservation 142
MWCNT 105

nanofluid 85
nanomaterials 79
Navier–Stokes equations 82
no-slip boundary conditions 83
Nusselt number 83

on-grid 49
organic electrolyte 186, 188, 197
outlet 103–112
oxygen evolution reaction 8–9
oxygen gas 150, 151, 152, 153, 154, 155, 156, 157

PEM fuel cell 135, 138
photocatalyst 3
   charge transfer 18
   mechanism 3–7
photon energy 44
photovoltaic 43, 45
physicochemical properties 186, 187
power density 172
pressure-based solver 145
pressure contours 151, 154, 157
PV system 29
   benefits 29
   challenges 29

quaternary 57, 66, 69–70

radiation 101–103, 106–112
Rayleigh number 83
renewable energy 43
residual iterations 147

Schottky Junction 15–17
semiconductor heterostructure 15
separator 173
solar cell 44, 48, 57–73, 186, 196
   dye-sensitized solar cell 196
   organic solar cells 199
   perovskite solar cells 198
solar PV array 48
solar radiation 45
solar radiation intensity 103, 105, 109
solar spectrum 43
solid electrolyte interface 187, 189
solution controls 146
specific energy 176
streamline contours 152, 155, 158
streamlines 88, 89, 90, 91
surface heat transfer coefficient 92

temperature contours 150, 153, 156
ternary 57, 65–66, 68
thin-film 56, 58–59, 60–69, 71–72
Tiwari–Das model 84
total heat flux 92
trapezium 82

U-tube 101–107, 109, 112
validation 165

VASP 67
volume fraction 94
vortex 162, 163, 164
vortex cells 86

waste to wealth 120
water heater 101–103, 105–107, 109–112
Wien2k 67–68

zinc 94, 95
Z-scheme 15, 16, 17